Ship Construction Sketches & Notes

SHIP CONSTRUCTION SKETCHES & NOTES

Second Edition

Kemp & Young
Revised by David J. Eyres

BUTTERWORTH HEINEMANN

OXFORD   AMSTERDAM   BOSTON   LONDON   NEW YORK   PARIS
SAN DIEGO   SAN FRANCISCO   SINGAPORE   SYDNEY   TOKYO

Butterworth-Heinemann
An imprint of Elsevier Science
Linacre House, Jordan Hill, Oxford OX2 8DP
225 Wildwood Avenue, Woburn MA 01801-2041

First published by Stanford Maritime Ltd 1968
Second edition 1997
Reprinted 1999 (twice), 2000
Transferred to digital printing 2002

British Library Cataloguing in Publication Data
A catalogue record for this book is available from the British Library

Library of Congress Cataloguing in Publication Data
A catalogue record for this book is available from the Library of Congress

ISBN 0 7506 3756 0

For information on all Butterworth-Heinemann Publications
visit our website at www.bh.com

Printed and bound in Great Britain by Antony Rowe Ltd, Eastbourne

Preface

Ship Construction Sketches & Notes has been popular for many years with Merchant Navy Officers studying for their statutory examinations. This edition updates the subject and presents the content in a more appropriate sequence whilst, wherever possible, retaining the sketches and concise notes which students have found helpful.

David J. Eyres
1997

Illustrations

Ship dimensions and terms

The ship's size and its form may be defined by a number of dimensions and terms.

LENGTH OVERALL is the length of the ship taken over all extremities.

LENGTH BETWEEN PERPENDICULARS is the length between the aft and forward perpendiculars measured along the summer load line.

AFTER PERPENDICULAR is a perpendicular drawn at the point where the aft side of the rudder post meets the summer load waterline. Where no rudder post is fitted it is taken as the centreline of the rudder stock.

FORWARD PERPENDICULAR is a perpendicular drawn at the point where the foreside of the stem meets the summer load line.

MIDSHIPS is a point midway between the after and forward perpendiculars.

Where moulded dimensions are referred to these are taken to the inside of the plating on a ship with a metal hull.

MOULDED BEAM is measured at midships and is the maximum moulded breadth of the ship.

MOULDED DEPTH is measured at midships and is the depth from the base line to the underside of the deck at the ship's side.

MOULDED DRAUGHT is measured at midships and is the depth from the base line to the summer load line.

BASE LINE is a horizontal line drawn at the top of the keel plate.

LIGHT DISPLACEMENT is the weight of the hull, engines, spare parts, and with water in the boilers and condensers to working level.

LOAD DISPLACEMENT is the weight of the hull and everything on board when floating at the designed summer draught.

DEADWEIGHT CARRYING CAPACITY is the difference between the light and loaded displacements and is the weight of cargo, stores, ballast, fresh water, fuel oil, crew, passengers and effects on board.

STATUTORY FREEBOARD is the distance from the upper edge of the summer load line to the upper edge of the deck line.

RESERVE BUOYANCY is virtually the (available) watertight volume above the waterline.

SHEER may be defined as the rise of a ship's deck fore and aft. It adds buoyancy to the ends where it is most needed. A correction for non-standard sheer is applied when calculating the freeboard.

CAMBER OR ROUND OF BEAM is the curvature of the decks in the transverse direction, measured as the height of deck at the centreline above the height of deck at side. It helps to shed water from the deck and adds to its longitudinal strength.

FLARE is the outward curvature of the side shell above the waterline at the forward end of the ship. It increases buoyancy thus limiting sinkage of the bow into head seas, promotes dryness forward and provides a wider forecastle deck allowing the anchors to drop clear of the shell plating.

TUMBLEHOME is the inward curvature of the side shell above the waterline. Modern ships rarely have tumblehome.

RISE OF FLOOR is the rise of the bottom shell plating above the horizontal base line, measured at the ship's side. The object is to provide for the drainage of liquids to the ship's centreline.

Many of these terms and others, which are self explanatory, are illustrated (below).

DIMENSIONS

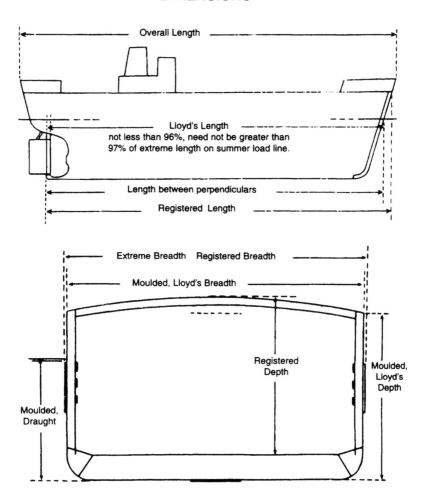

Classification

The principal maritime nations have Classification Societies whose primary function is to survey ships so as to assess the adequacy of their strength and construction, and for which purpose they publish rules. The British Classification Society is Lloyds Register of Shipping which classes most British shipping and, as it has world-wide connections with surveyors in the principal ports, a significant proportion of the world's tonnage. The scantlings (sizes) of the materials to be used, as well as certain items of equipment (anchors, cables and warps), can be obtained from Lloyds, 'Rules and Regulations for the Classification of Ships'. This publication is amended and updated on a regular basis. The scantlings are based on the basic dimensions of the ship shown [on page 5] and defined below, detailed calculations of the still water bending moment and the section modulus of the particular item in association with other structural members.

Length L is the distance in metres on the summer load waterline from the foreside of the stem to the after side of the rudder post or to the centre of the rudder stock if there is no rudder post. L is not to be less than 96% of extreme length on summer load waterline and need not be more than 97% of that length.

Breadth B is the greatest moulded breadth in metres.

Depth D is measured in metres at the middle of length L from the top of the keel to the top of the deck beam at side on the uppermost continuous deck. With a rounded gunwale D is measured to the continuation of the moulded deck line.

Draught d is the moulded draught in metres.

A ship built to Lloyds highest class will be given this character,
+ 100 A 1
+ indicates 'built under survey' which means the plans were submitted and approved, all steel was manufactured at an approved steelworks, and the construction was overseen by a surveyor.
100A indicates the scantlings are in accordance with the Rules.
1 indicates the equipment is in accordance with the Rules.

Ships built for a particular type of service have a Class Notation in addition to the above, e.g.100A 1 Liquified Gas Carrier

If the ship's machinery is built and installed under Lloyds survey the character L.M.C. (Lloyd's Machinery Certificate) is assigned.

Where additional strengthening is fitted for navigation in ice an appropriate notation may be assigned. The notations fall into two categories: those for 'first-year ice' i.e. where waters ice up in winter only; and 'multi-year ice' i.e. Arctic and Antarctic waters. The latter includes the term 'icebreaker' in the class notation.

TERMS

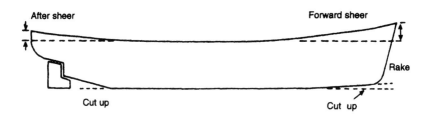

After sheer

Forward sheer

Rake

Cut up

Cut up

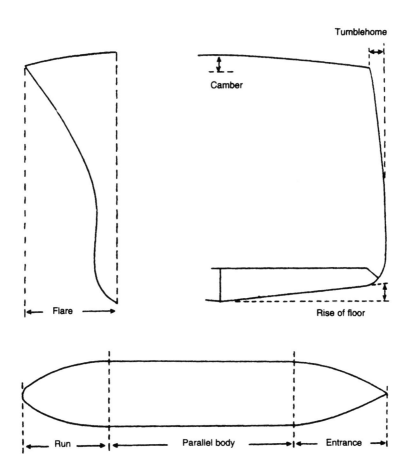

Tumblehome

Camber

Flare

Rise of floor

Run

Parallel body

Entrance

In order to remain in class a classified ship is required to undergo surveys at regular intervals, as follows:

(a) Annual, (b) Intermediate (instead of 2nd or 3rd annual), (c) Special (every 5 years) or Continuous surveys where a maximum period of five years is allowed between the consecutive examination of each part. Special surveys increase in severity as the vessel gets older. Ships are to be examined in dry-dock at periods coinciding with the special and intermediate survey. An 'in-water' survey in lieu of the intermediate survey docking may be accepted.

All damage must be surveyed and repaired to the satisfaction of the Society's Surveyor.

A ship which is not classified will still have to reach a minimum standard of strength and have similar surveys. All [larger] ships trading internationally are required to have a 'Ship Safety Construction Certificate' issued by, or on behalf of, the government of the country of registration.

A large percentage of maritime insurance is effected at Lloyd's of London. Although the name is the same as that of the Classification Society there is no direct connection and the two should not be confused.

International shipping conventions

International shipping is regulated by conventions, the requirements of which are agreed at conferences convened by a United Nations agency, the International Maritime Organisation (IMO). These conventions come into force when a stipulated number of countries, that are members of IMO, become party to the convention by applying its requirements to their national shipping.

The following international conventions have a significant influence on ship design and construction.

International Convention on Load Lines of Ships, 1966
International Convention on Tonnage Measurement, 1969
International Convention for the Safety of Life at Sea, 1974 (SOLAS 74)
International Convention for the Prevention of Pollution from Ships, 1973 and its Protocol of 1978 (MARPOL 73/78)

Load Lines

Under the International Convention on Load Lines all ships which are 24 m or more in length, except ships of war, fishing boats and pleasure boats, must have a load line. Such ships trading internationally are marked with a load line assigned by the maritime authority of the flag state or a Classification Society authorized by the flag state. The initial letters of the assigning authority are cut in on each side of the load line disc. For the United Kingdom, the Marine Safety Agency is the maritime administration. National legislation may also require ships which do not trade internationally to be assigned and marked with a load line to which they may be safely loaded. Shown [below] are a full set of markings assigned under the International Convention on Load Lines which include the zone, seasonal and freshwater allowance markings. On the left are shown the additional freeboard markings assigned to a ship carrying timber deck cargoes. On the assumption that the timber cargo provides additional buoyancy it will be noted that the ship may load to a deeper draught except in the case of the Winter North Atlantic (WNA) zone.

LOAD LINES

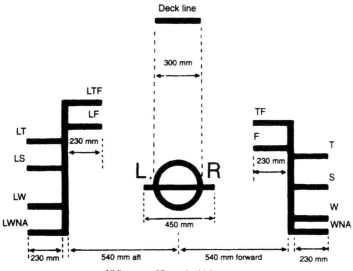

Tonnage

Tonnage is a measure of the cubic capacity of a ship. The gross tonnage of a ship is indicative of the total volume of the enclosed spaces of a ship and may often be used in reference to the size of the ship. Net tonnage is indicative of the volume of the cargo and passenger spaces in a ship which produce the revenue. Most charges levied on a ship are based on its tonnages.

Measurement of a ship for tonnage is undertaken by the maritime authority of the flag state or a Classification Society authorized by the flag state. A universal system of measurement for tonnage has been established under the International Convention on Tonnage Measurement of Ships 1969. Ships measured in accordance with this convention are issued with an International Tonnage Certificate which indicates the ships gross and net tonnages and is accepted in ports worldwide.

Gross tonnage

The gross tonnage (GT) is determined by the formula: $GT = K_1 V$
where $K_1 = 0.2 + 0.02 \log_{10} V$
and V = the total volume of all enclosed spaces in cubic metres.

Net tonnage

The net tonnage (NT) is determined by the following formulae:
(1) For ships carrying more than 12 passengers $NT = K_2 V_c \left[\dfrac{4d}{3D}\right]^2 + K_3 N_1 + \dfrac{N_2}{10}$

(2) For other ships $NT = K_2 V_c \left[\dfrac{4d}{3D}\right]^2$

where
V = total volume of cargo spaces in cubic metres.
d = moulded draft amidships in metres (summer load line draught or deepest subdivision load line in case of passenger ships).
D = moulded depth in metres amidships.
K_1 = $0.2 + 0.02 \log_{10} V$.
K_2 = $1.25 \dfrac{(GT + 10\,000)}{10\,000}$.
N_1 = number of passengers in cabins with more than 8 berths.
N_2 = number of other passengers.
$N_1 + N_2$ = total number of passengers the ship is permitted to carry.
The factor $\left[\dfrac{4d}{3D}\right]^2$ is not to be greater than unity.
The term $K_2 V_c \left[\dfrac{4d}{3D}\right]^2$ is not to be taken as less than $0.25GT$.
Net Tonnage is not to be taken as less than $0.3GT$.

Oil tankers with segregated ballast tanks may have these tanks measured separately and the tanker's International Tonnage Certificate can indicate the ship's gross tonnage with these spaces deducted. This is to promote the provision of segregated ballast tanks and protection of the cargo tanks.

Suez Canal and Panama Canal tonnages

Tolls for passage of the Suez and Panama Canals are based on a tonnage measurement of the ship. The Panama Canal tonnage measurement system is now compatible with the universal measurement system described above, but the Suez Canal tonnage measurement rules pre-date the universal measurement system.

SOLAS 74

The International Convention for the Safety of Life at Sea, 1974 includes standards relating to the intact and damage stability of ships, sub-division, machinery and electrical installations, structural fire protection, carriage of grain and dangerous goods, all of which have a significant influence on the design and construction of ships.

MARPOL 73/78

Substantial requirements relating to the design and construction of oil tankers are contained in the International Convention for the Prevention of Pollution from Ships, 1973, and particularly its Protocol of 1978. These requirements aimed at minimizing outflows of oil include limitations on cargo tank size, provision of clean and segregated ballast tank spaces, protection of the cargo tank spaces by double hull structures etc. For dry cargo ships MARPOL 73/78 prohibits the carriage of oil fuel in the forepeak and use of oil fuel tanks for carriage of water ballast. Detailed requirements concerning the construction of chemical carriers and other ships carrying noxious liquid substances are also covered by this convention.

Strength of materials

When a force, or a load, is applied to a solid body it tends to change the shape of the body. When the applied force is removed the body will regain its original shape. The property, which most substances possess, of returning to their original shape is termed 'elasticity'.

Should the applied force be large enough, the resistance offered by the material will be overcome and when the force is removed the body will no longer return to its original shape and will have become permanently distorted.

The point at which a body ceases to be elastic and becomes permanently distorted is termed the 'yield point' and the load which is applied to cause this is the 'yield point load'. The body is then said to have undergone 'plastic deformation or flow'. Whenever a change of dimensions of a body occurs a state of strain is set up in that body.

Stress and strain

Stress is a load or force acting per unit area and may be expressed in kilogrammes per square millimetre (kg/mm^2). Stresses are of three main types:

1) Tensile   forces acting in such a direction as to increase the length.
2) Compressive forces acting in such a direction as to decrease the length.
3) Shear   the effect of two forces acting in opposite directions and along parallel lines. The forces act in such a direction so as to cause the various parts of a section to slide one on the other.

Stress is proportional to the distance from the neutral axis of the body to which the force is applied. The neutral axis passes through the centroid (geometric centre) of the body.

Strain is the distortion in a material due to stress.

See illustrations and 'stress–strain' curve on pages 13 and 15.

Mechanical properties of metals

Plasticity	The ease with which a metal may be bent or moulded into a given shape.
Brittleness	The opposite of plasticity, lack of elasticity.
Malleability	The property possesed by a metal of becoming permanently flattened or stretched.
Hardness	The property of a metal to resist wear and abrasion.
Fatigue	A metal subjected to continually applied loads may eventually fail from fatigue.
Ductility	Ability to be drawn out lengthwise, the amount of the extension measures the ductility.

STRESSES

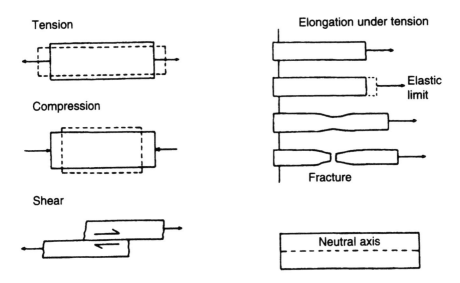

Tension

Compression

Shear

Elongation under tension

Elastic
limit

Fracture

Neutral axis

Brittle fracture

When a tensile test is applied to a metal it elongates elastically, then plastically and finally fractures. A warning of impending fracture is given by the preceding elongation.

Occasionally mild steel behaves in a completely brittle manner. The fracture occurs without warning at a stress well below the elastic limit of the mild steel. A fracture of this nature is known as 'brittle fracture'. The resulting crack may travel at a very high speed (up to 2000 m/s).

Factors related to the occurrence of brittle fracture are (a) the presence of a tensile stress, (b) the metallurgical properties of the mild steel, and (c) presence of a defect or poor structural design detail which provides a 'notch' from which the crack is initiated. Usually brittle fracture occurs at a relatively low temperature. Thicker plate is more prone to brittle fracture.

Whilst brittle fracture can occur in both welded and riveted structures its effects were more noticeable with the advent of larger all-welded ship structures. Given that welded plates are continuous a brittle fracture crack may travel through the structure unhindered. Larger ship structures are required to have mild steel plate with metallurgical properties which are less prone to brittle fracture at strategic locations and where thicker plate is used. Particular care is also exercised with the quality of welding and structural design detail to avoid defects which may initiate brittle fracture. The property of a mild steel which makes it less prone to brittle fracture than another mild steel is its greater 'notch toughness'.

Measurement of sectional strength

A beam when loaded tends to bend, and the amount it bends or deflects from the normal under that load is determined by the beam's resistance to bending. The resistance to bending is a function of the strength of the material from which the beam is constructed and the geometry of its cross-section. The factor which relates to the geometrical form of its cross-section is termed the 'moment of inertia' of the beam.

The moment of inertia I is a measure of a beam's ability to resist deflection; it is an indication of how the mass is distributed with respect to the neutral axis. With a given cross-sectional area it is possible to create a number of different sections. One section will have a greater I than another because of the greater distances of its flanges from the neutral axis.

STRESSES

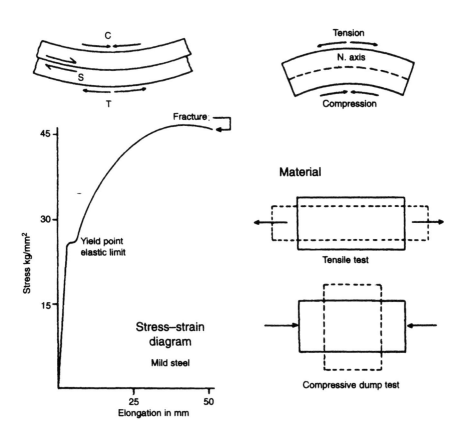

The distance of the upper (or lower) flange from the neutral axis (designated by y), is an indication of the efficiency with which the flange can resist stresses due to bending.

If the moment of inertia I is divided by y the resultant expression I/y can be used as a standard or modulus of the ability of a section to withstand bending and associated stresses. The expression I/y is called the section modulus.

Although the geometrical distribution of material in its cross-section is a measure of the strength of the beam, the material used also determines the strength of the beam. The greater the strength of the material the greater will be the beam's resistance to bending.

Resistance to bending implies stress p, the maximum stress occurring at the uppermost and lowermost parts of the loaded beam.

$$\text{Total resistance to bending} = \frac{I}{y} \times p.$$

$$\text{Moment tending to bend beam} = M$$

$$\text{then } M = \frac{I}{y} \times p.$$

If the bending moment M and maximum stress p which can be permitted (due to material) are known,

$$\text{then } \frac{M}{p} = \frac{I}{y} = \text{section modulus.}$$

It is then necessary to select a section with a sectional modulus at least equal to that required. Reference may be made to published 'Geometrical Properties of Rolled Sections and Built Girders' for this purpose.

In obtaining the scantlings of the ship's various structural items the section modulus of the section, in association with the plate or other structural item to which it is attached, has to be calculated.

SECTION MODULUS

Sections of equal area 6000 mm²

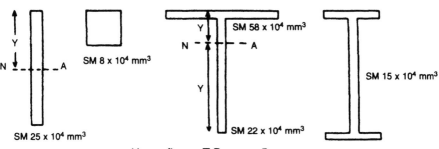

SM 25 x 10⁴ mm³

SM 8 x 10⁴ mm³

SM 58 x 10⁴ mm³

SM 22 x 10⁴ mm³

SM 15 x 10⁴ mm³

Upper flange T Bar max 5 m

Forces to which a ship is subjected

A ship at sea is subjected to a number of forces causing the structure to distort. These may be divided into two categories, as follows:

(1) Static forces	Ship floating at rest in still water. Two forces acting, (a) weight of ship acting vertically downward and (b) water pressure acting perpendicular to outside surface of hull.
(2) Dynamic forces	Ship in motion. Six ship motions are illustrated on page 19. When these motions are large then very large forces may be generated . These forces are often of a local nature, e.g. heavy pitching resulting in pounding forward, but they are liable to cause the structure to vibrate and thus transmit the stresses to other parts of the structure.

FORCES CAUSING STRESS

Static forces

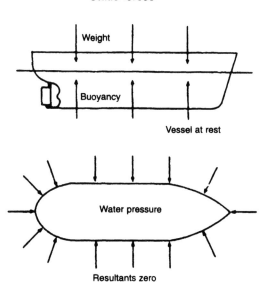

Dynamic forces

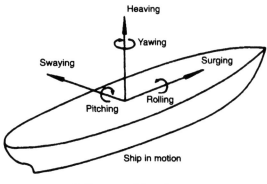

6 degrees of freedom

These forces produce stresses in the ship's structure which may be divided into two categories, as follows:

(1) Global   those affecting the whole ship.

(2) Local   those affecting a particular part of the ship.

Global stresses

(a)   Longitudinal stresses in still water.
Although the upthrust (buoyancy) is equal to the weight of the ship the distribution of weight and buoyancy is not uniform throughout the length of the ship and differences (load) occur throughout the length, giving rise to tensile and compressive stresses away from the neutral axis.

(b)   Longitudinal stresses in a seaway causing hogging and sagging.
When the ship is amongst waves the weight distribution remains unchanged but the distribution of buoyancy is altered.

(c)   Shearing stresses.
The longitudinal stresses imposed by the weight and buoyancy distribution can give rise to longitudinal shearing stresses. The maximum longitudinal shearing stress occurs at the neutral axis and decreases to a minimum at the deck and keel. Vertical shearing stresses also occur as the result of the non-uniform longitudinal distribution of weight and buoyancy.

STILL WATER CONDITIONS

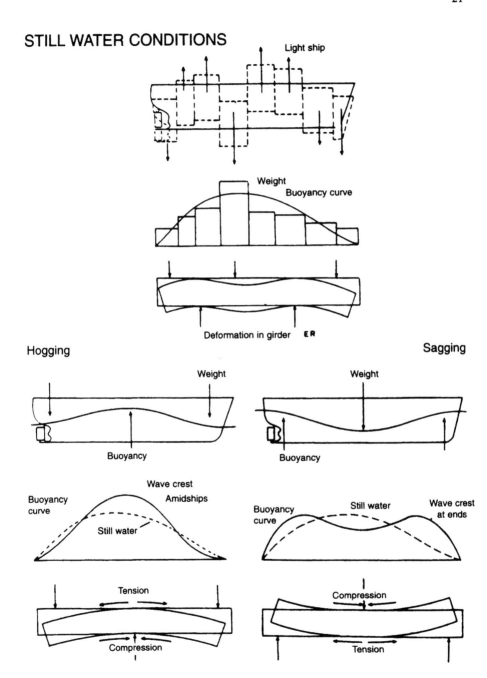

Light ship

Weight
Buoyancy curve

Deformation in girder   E R

Hogging

Weight

Buoyancy

Sagging

Weight

Buoyancy

Buoyancy
curve

Wave crest
Amidships

Still water

Buoyancy
curve

Still water

Wave crest
at ends

Tension

Compression

Compression

Tension

Other forces which produce global stresses are:

(d) Racking.
When a ship is rolling the accelerations on the ship's structure are liable to cause distortion in the transverse section. The greatest effect is under light ship conditions.

(e) Torsion.
A ship traversing a wave train at an angle will be subject to righting moments of opposite directions at its ends. The hull is subject to a twisting moment (torque) and the structure is in 'torsion' (see page 25). The greatest effect occurs with decks having large openings.

(e) Water pressure.
Water pressure acts perpendicular to the surface and increases with depth. The effect of water pressure is to push in the ship's sides and push up the ship's bottom.

(f) Drydocking.
In drydock there is a tendency to set up the ship's keel due to the upthrust of the supporting keel blocks resulting in a change in the shape of the transverse section.

Local stresses

(a) Panting (see page 52)

(b) Pounding (see page 46)

(c) Local loading
Localized heavy weights e.g machinery, or localized loading of heavy cargo e.g. ore may give rise to localized distortion of the transverse section.

(d) The ends of superstructures
These may represent major discontinuities in the ship's structure giving rise to localized stresses which may result in cracking.

(e) Deck openings
Holes cut in the deck plating, e.g. hatchways, masts, etc. create areas of high local stress due to the lack of continuity (of structure) created by the opening.

23

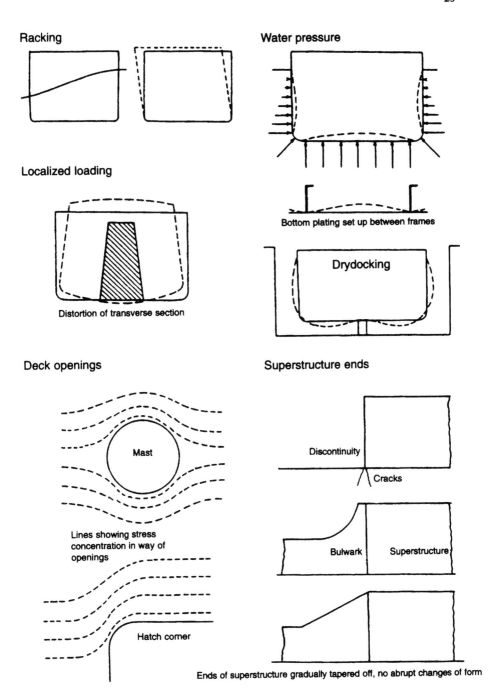

Racking

Water pressure

Bottom plating set up between frames

Drydocking

Localized loading

Distortion of transverse section

Deck openings

Mast

Lines showing stress concentration in way of openings

Hatch corner

Superstructure ends

Discontinuity

Cracks

Bulwark

Superstructure

Ends of superstructure gradually tapered off, no abrupt changes of form

(f)   Other examples where local stresses may occur:

Vibration due to propellers.
Stresses set up by stays, shrouds etc.
Stresses set up in the vicinity of hawse pipes, windlass, winches etc.

The materials used in a ship's structure form a box-shaped girder of very large dimensions.

The side shell plating, keel and bottom plating, deck plating, hatch coamings, deck girders, double bottom structure, bottom, side and deck longitudinals and any longitudinal bulkheads assist in overcoming longitudinal stresses. Transverse bulkheads and deep transverses are efficient in preserving the transverse form. Frames, beams and floors etc. all being securely bracketed together help to stiffen the plating against compressive stresses. Since water pressure is a major stress on the hull, and increases with depth, the bottom plating is heavier and the side framing size reduces with height above the bottom.

It is essential to prevent the various stresses causing deformation or possible fracture of the structure. This may be achieved by increasing the sizes of material used, by careful disposition of the material and by paying careful attention to the structural design detail.

TORSION

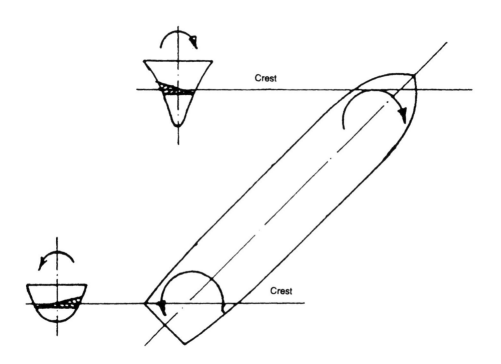

Crest

Crest

Curves of shearing force and bending moment

The shearing force and bending moments at sections along the length of a beam may be shown graphically by plotting the values of the shearing force and bending moment at various points along the beam. The resultant graphs may be straight lines or curves and show the variation of stress along the beam.

Such curves are explained in the companion volume *Ship Stability Notes and Examples* by Kemp and Young.

The illustrations opposite show the associated curves of shear force and bending moments for a ship in the still water condition and when amongst waves. In the first instance the wave crest is amidships and the troughs are at each end of the ship and in the second instance the wave crests are at each end and the trough amidships, i.e. 'hogging' and 'sagging'.

TYPICAL STRENGTH CURVES

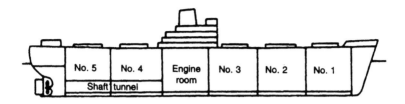

Still water condition

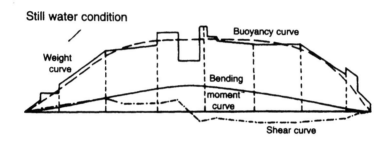

Hogging condition

Sagging condition

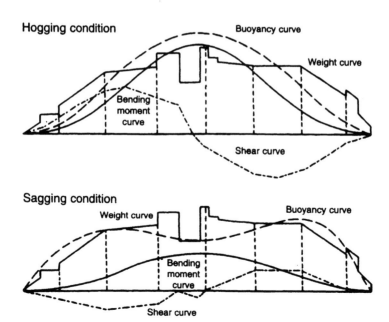

Steel

The production of steel starts with the smelting of iron ore and the making of pig-iron. Pig-iron is from 92–97% iron, the remainder being carbon, silicon, manganese, sulphur and phosphorus. In the subsequent manufacture of steels the pig-iron is refined, in other words the non-metallic content is reduced.

Steels may be broadly considered as alloys of iron and carbon, the carbon percentage varying from about 0.1–0.2% in mild steels to about 1.8% in some hardened steels.

The properties of steels may be altered greatly by the heat treatment to which they are subjected after initial production. These heat treatments bring about a change in the mechanical properties principally by modifying the steel's structure.

Shipbuilding steels are made by the open hearth, electric furnace or oxygen process and can be subject to heat treatments such as annealing, normalizing, quenching and tempering.

Mild steel with a carbon content of 0.15–0.23%, reasonably high manganese content and a minimum of sulphur and phosphorus, is primarily used for welded ship construction purposes. It is relatively cheap and may be reasonably easily worked without any appreciable loss of its mechanical properties. It lacks notch toughness and is subject to brittle fracture, particularly in thicker plate.

There are five grades of steel, A to E, used in shipbuilding; the grades varying according to the alloying elements. Grades A and B are ordinary mild steels, grades C, D and E possess higher notch toughness characteristics.

The classification society rules specify the grade to be used, where thicker plate and notch toughness is desirable. For example, in ships of 250 m or less in length the sheerstrake over 40% of the length amidships is to be; Grade A if less than 15 mm, Grade B 15–20 mm, Grade D 25–40 mm and Grade E where over 40 mm.

In large oil tankers, ore carriers etc., high-tensile steels are used. These steels with a higher-strength than mild steel permit a reduction in thickness of the deck, bottom shell and longitudinal framing over the midships portion of the hull and a subsequent saving in weight. Because of the lesser thickness the hull may be subject to greater deflections and the effects of corrosion require more vigilant inspection. Higher-tensile steels are graded according to both their strength level and notch toughness.

Both mild steel and higher-tensile steel plates and sections built into a ship are to be produced at works approved by the classification society. During production, an analysis of the steel is required and so are specified tests of the rolled steel to ensure its compliance with the rules. Tensile and impact tests are made on specimens obtained from the same product as the finished steel plate or section.

Rolled and built sections

Various rolled steel sections and built sections are used in ship construction and are illustrated. The type of section used depends on the degree of strength required and often the depth of web that can be accommodated. Built sections are used when a greater degree of strength is required than that obtained from available rolled sections.

ROLLED SECTIONS

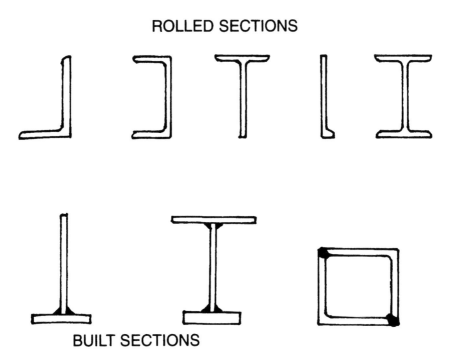

BUILT SECTIONS

Aluminium alloy

Aluminium alloys are presently used in shipbuilding for the superstructures of certain larger ships and the construction of many high-speed craft.

The advantage of using aluminium alloy, especially for superstructures and high-speed craft, is its light weight in relation to its strength. Aluminium alloy superstructures reduce topweight and therefore improve stability and can also allow an increase in deadweight. There are disadvantages in that the material is more expensive than mild steel and, as it has a low melting point, it is unsuitable where a higher standard of fire protection is required

A useful property of aluminium alloy is its high resistance to corrosion. This is due to the oxide film which forms on its surface and previously presented problems in welding the material which has been overcome with the use of inert-gas shielded arc welding. However, corrosion of the aluminium alloy can occur through galvanic action when it is in contact with another metal in the presence of an electrolyte (e.g. sea water). Exposed bi-metallic connections are therefore insulated to avoid any aluminium alloy to steel contact. Where the sides of an aluminium alloy superstructure are attached to the steel deck special attention is paid to the connection. The two metals are separated by using a non-absorbent material, e.g. Neoprene tape. The aluminium alloy house side is lapped on the outside of the steel coaming to prevent water collecting at the interface. The joint may be riveted or bolted with galvanized steel bolts, but modern practice is to make use of an explosion bonded alloy/steel transition joint to which the aluminium alloy house side and steel coaming are directly welded

Pure aluminium is of little use for structural purposes and is therefore alloyed with other materials, cold worked and/or heat treated to improve its tensile strength. Specifications for aluminium alloys which are suitable for shipbuilding are found in the rules of classification societies such as Lloyds Register of Shipping.
In the preservation of aluminium alloys LEAD BASED paints should NEVER be used.

Plastics

There is a wide and varied use of plastics in shipbuilding. These are used in many different forms to exploit the variety of properties of plastics. Useful properties can be lightweight, flexibile, durable, not highly inflammable, good insulation, and ease of fabrication.

A few of the uses for which plastics may be used on board a ship are given below:
Fibre – reinforced plastic (FRP) for the construction of lifeboats.
Laminated plastic bearings – stern tube.
Nylon, terylene – mooring ropes.
Pvc etc. – non-essential piping systems.
Fibreglass – insulation.
Electrical cable insulation.
Decorative laminates for accommodation linings.

ALUMINIUM ALLOY TO STEEL CONNECTION

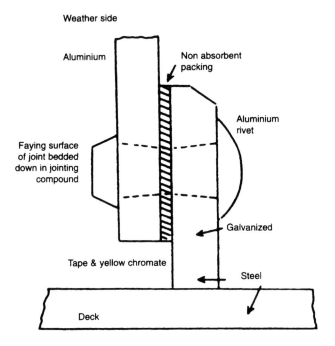

Weather side

Aluminium

Non absorbent packing

Aluminium rivet

Faying surface of joint bedded down in jointing compound

Galvanized

Tape & yellow chromate

Steel

Deck

Welding and cutting

The welding processes used in shipbuilding are of the fusion welding type. Fusion welding is achieved by means of a heat source which is intense enough to melt the edges of the material to be joined as it is traversed along the joint.

The heat source may be generated in a number of ways, examples are:

(a) Gas welding
A gas flame was probably the first heat source for fusion welding, and the most commonly used gas is acetylene, which with oxygen produces the high temperature flame. Oxy-acetylene welding is only really suitable for thinner mild steel plate and its use in shipbuilding is limited to the fabrication of sheet metal items like ventilation trunking, cable trays etc. and some plumbing work.

(b) Electric arc welding
An electric arc is formed when an electric current passes between two electrodes separated by a short distance. In electric arc welding one electrode is the welding rod or wire while the other is the metal to be welded. The welding electrode and the plate are connected to the electrical supply and a high temperature arc is created by momentarily touching the electrode onto the metal and then withdrawing it to create a small gap between it and the metal. The arc will melt the edges of the metal joint and the consumable welding rod or wire.

Consumable manual welding rods have flux coatings which provide inert gas shielding for the arc and molten metal. The gas shielding consumes the surrounding atmospheric gases which might otherwise be absorbed by the molten metal, stabilizes the arc, and provides a protective slag for the molten metal. Automatic welding processes may employ consumable wires with an external or cored flux which serves the same purpose. Submerged arc welding where the arc is maintained within a blanket of granulated flux is commonly used for downhand automatic welding of steel in shipbuilding.

Inert gas shielded arc welding is used for welding aluminium alloys, usually with argon as the gas, and using a tungsten electrode for manual welding of light plate or consumable metal wire for semi-automatic or automatic welding of heavier plate. Mild steel inert gas shielded welding is now also common using automatic or semi-automatic processes with CO_2 as the shielding gas.

(c) Thermit welding
A combination of chemicals called the thermit is fired producing a chemical reaction. It is essentially a casting process and this method is mainly used in the joining of steel castings e.g. sternframes.

ELECTRIC ARC WELDING

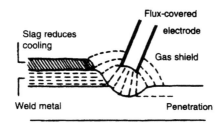

Good weld

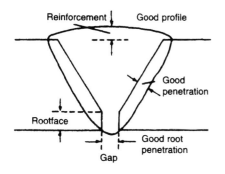

Faults

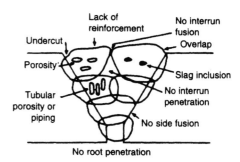

Welding practice

In joining two plates (in the same plane) a stronger and more efficient joint is obtained by 'butt' welding rather than 'lap' welding the plates. Where the plate thickness exceeds 5–6 mm it is necessary to make more than one welding pass to deposit sufficient weld metal to close the butt joint. It is necessary to bevel the edges of thicker plates in order to achieve complete penetration of weld metal (see Cutting). Where thicker insert plates are butt welded to the thinner surrounding plate the heavier insert plate is chamfered to the thickness of the adjacent plate before the butt edge bevel is cut. To ensure complete penetration of a butt weld it is necessary to turn the plates and complete a 'back run' weld unless a backing bar or special 'one-sided' welding technique is used. 'One-sided' butt welding techniques are common in ship fabrication where a minimum of plate and unit turning results in time and cost savings.

'Fillet' welds are used to attach sections to plate or one plate perpendicular to another. Fillet welds may be continuous or intermittent depending on the structural effectiveness of the member to be welded. Where fillet welds are intermittent they may be either staggered or chain welded (see figure on page 35), the section may also be scalloped to give the same result when continuously welded.

It is desirable that a maximum of welds are made in the downhand position where the ease of depositing weld metal and the common use of fully automatic welding results in higher quality welds. For this reason many ship units are fabricated upside down e.g. the deck plating being assembled and welded and then the deck beams or longitudinals, girders and transverses being welded on top of the plating before the whole unit is lifted and turned for erection. Vertical welding of side shell units is necessary and is normally accomplished by working upwards. Automatic welding processes are available for this purpose. Overhead welding is the most difficult and requires skill and special techniques when carried out manually or with semi-automatic equipment.

Welding sequence

During welding heat is applied to the plate which will expand locally and on cooling contract. This can lead to distortion of the structure or result in residual stresses where restraint is applied to limit distortion. To reduce distortion and minimize residual stresses it is important that the correct welding sequence is followed throughout the construction. In welding the side shell plating of a ship for example the butts are welded first and then the adjacent seams working outwards from the centre both vertically and longitudinally. Stiffening members are left unwelded for a distance across the plate butts and seams and when finally welded are notched or scalloped in the way of the seam or butt. In repair work correct welding sequences are also important, particularly where new material is to be inserted into the existing relatively rigid structure. Existing seams, butts and welds of stiffening members will need to be cut back some distance and re-welded in sequence with the new insert.

Sections used in welded construction

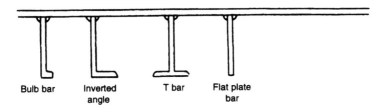

Bulb bar   Inverted angle   T bar   Flat plate bar

Butt weld

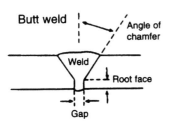

Angle of chamfer

Weld

Root face

Gap

Staggered intermittent welding

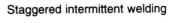

d s

s not less than 75 mm

Chain intermittent welding

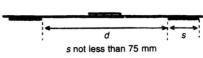

d s 150 mm max

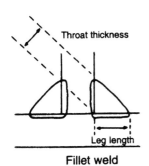

Throat thickness

Leg length

Fillet weld

Radius not less than 25 mm D Depth not greater than 0.25 D or 75 mm whichever is less

s 150 mm max

d

Scalloped frames, longitudinals with double fillet welds d given in rules

Single V fillet

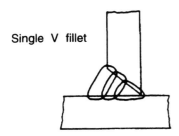

Weld testing

Welding materials are subjected to comprehensive tests before they are approved by a classification society for use in ship work. Most testing of finished welds in shipbuilding is by visual inspection. Spot checks and examination of critical welds in ship structures are undertaken using radiographic or ultrasonic equipment.

Various faults may be observed in butt and fillet welds and are due to one or more factors, e.g. bad design, incorrect welding procedure, use of wrong materials, bad workmanship etc. Different faults are illustrated on page 33.

Cutting

The cutting to shape and edge preparation of plates is now a highly automated process in shipbuilding. Gas cutting is the most commonly used method with acetylene being used with oxygen to provide the flame for preheating the mild steel. Once the metal is preheated a confined stream of oxygen is introduced which oxidizes the iron in a narrow band and the molten oxide and metal is removed by the force of the oxygen stream. A narrow parallel-sided gap is left between the cut edges. Plasma cutting is now used in many shipyards and can be used to cut all electrically conductive materials. Laser cutting is also being utilized in modern shipyards.

To achieve the bevelled plate edges required for multi-run welds in thicker plates an oxy-acetylene cutting machine may be fitted with more than one nozzle to achieve the desired cut bevels in one pass (see figure on page 37).

PLATE EDGE PREPARATION

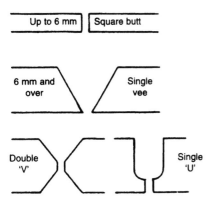

Up to 6 mm | Square butt

6 mm and over | Single vee

Double 'V' | Single 'U'

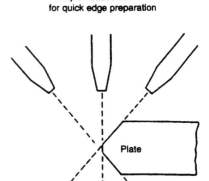

Triple headed burner
for quick edge preparation

Plate

Shipyard practice

Modern shipyard practice is based on computer-aided design and computer-aided manufacturing (CAD/CAM) systems which are readily available to industry.

Lines plan

The hull form of the ship is delineated on a scale drawing known as a Lines Plan. A set of Lines consists of three views as follows:

Profile or Sheer	– side elevation
Half Breadth Plan	– plan view
Body Plan	– cross-sectional view

A preliminary version of the Lines Plan will be prepared at the time of the conceptual design to give the required capacity, displacement etc. and is subsequently refined during the preliminary design stage to obtain the desired propulsive and seakeeping characteristics. The finished Lines Plan must be fair i.e. all the curved lines must run evenly and smoothly and there must be exact agreement between corresponding dimensions of the same point in the different views. When the small scale Lines Plan was manually drawn and faired the draughtsman would compile a 'table of offsets', i.e. a list of the half breadths at given heights above a base line to define each of the drawn cross-sections. These 'offsets' and the Lines Plan were then passed to loftsmen for full size or 10 to 1 scale fairing, or to a computer centre for full scale fairing. The loftsmen or computer centre would prepare a full set of faired offsets for each frame of the ship which would be utilized in its construction. With the use of integrated design systems on the shipyards computers, the conceptual creation of the hull form and its subsequent fairing for production purposes is accomplished without committing the plan to paper. The hull form is generally held in the computer system as a 3-dimensional 'wire model' which typically defines the moulded lines of all structural items so that any structural section of the ship can be generated automatically from the 'wire model'.

LINES PLAN

Sheer plan

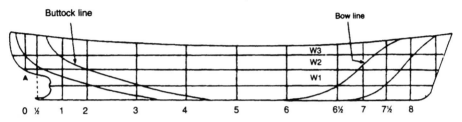

Half breadth plan

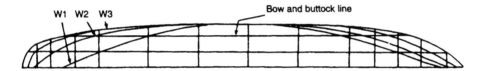

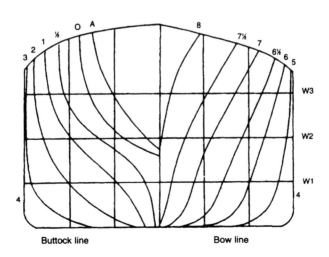

Body plan

Plate and section preparation

Initial preparation of the steel is essential to the efficiency of the shipbuilding process and the quality of the finished ship structure.

Before it is worked, steel plate is passed through a heavy set of straightening rolls known as mangles to ensure it is as flat as possible. The plate and steel sections are then shot blasted to remove rust and millscale before passing through an airless spray painting plant where a priming paint is applied. This protects the steel against rusting during the construction of the ship and prior to the application of any final paint coatings.

Plate cutting

Many of the plates in the ship's hull only require trimming and edge preparation (see page 37) and these are cut with a planing machine. Cutting is either by oxy-gas flame or plasma-arc, the cutting heads being mounted on beams which travel on rails along and across the table on which the plate lays. Plates which are to be cut to more intricate shapes with openings etc. are cut with a profiling machine which has some form of automatic control. In most shipyards numerical control is now used for plate profilers. Numerical control implies control of the machine by a tape on which is recorded the co-ordinate points of the desired plate profile. With the integrated software of CAD/CAM systems the plate profile and cutting data can be programmed and this data transferred to tape. The tape is read into a director which produces command signals to the servo-mechanism of the plate profiler.

Frame bending

To obtain the required curvature of rolled sections for transverse side framing and other curved members use is made of a hydraulically powered cold bending machine. In this machine the frame is progressively bent by application of a horizontal ram whilst the rolled section is held by gripping levers. The desired curvature can be obtained by marking on the straight length of rolled section the 'inverse curve' of that desired curvature. The length of rolled section reaches the desired curvature when it is bent to the extent that the marked 'inverse curve' becomes a straight line. The 'inverse curve' information can be determined using the CAD/CAM system, the frame line being defined by the computer-stored faired hull model. Also the cold bending machine may be numerically controlled with tapes defining the desired curvature.

Prefabrication

Construction of a ship follows a planned production process with fabrication of units under workshop conditions and their subsequent assembly on the building berth or dock. The 3-dimensional block assemblies which are taken to the berth or dock for erection may consist of a number of 2-dimensional sub-assemblies and can be outfitted with units of machinery, pipework and other ship systems prior to leaving the workshop. Each sub-assembly and block is designed to minimize positional welding and may be turned to facilitate this as well as the installation of outfit items in a block assembly. The size of these block assemblies are dictated by the available lifting capacities, dimensions that can be handled and the nature of the structure which must be self-supporting and accessible. Sequences of erection of the block assemblies vary but most often commence in the area of the aft machinery spaces where a significant amount of finishing work can still be required after erection. Sequential erection is from the bottom and upwards and fore and aft.

Launching

When the ship is built on a conventional berth and is almost ready for launching, launching ways will be set up. These consist of a fixed portion on the ground referred to as standing ways and a portion attached to the ship referred to as sliding ways. A lubricant is applied to the sloping standing ways to permit movement of the sliding ways over them. Shortly before the launch the weight of the ship is transferred from the building blocks to the launching ways and the ship is temporarily restrained to the moment of launch. Once in the water the ship is transferred to a berth for final fitting out and trials.

Modern shipyards often use a building dock, which may be under cover, in which the ship is erected. The ship is then launched by flooding the dock.

Where building berths are inadequate to cope with the dimensions of large tankers and bulk carriers, it is not uncommon to build the hull in two sections. The two parts may be launched, floated together, carefully aligned and welded together. If a large enough drydock is available the joining of the two parts may be more easily accomplished there.

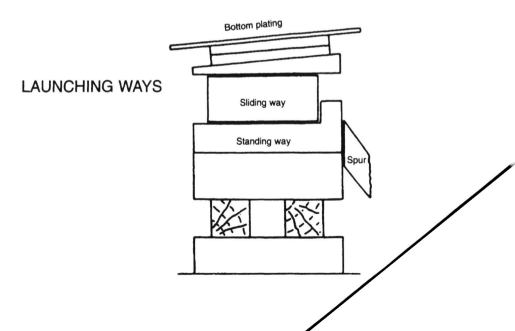

Bottom construction

The keel, centre girder and centreline strake of the tank top plating form a very strong I shaped girder which forms the 'backbone' of the ship.

The width and thickness of the keel strake is to be maintained over its whole length. In an average general cargo ship the width might be of the order of 1400 mm and thickness 20 mm (see page 48).

Double bottom construction

Framing within the double bottom is to be either longitudinal or transverse. The framing must be longitudinal in ships over 120 m in length and when the notation 'Heavy Cargoes' is assigned.

The thickness of the inner bottom plating is to be calculated and increased when the notation 'Heavy Cargoes' is assigned, or where there is no ceiling fitted in the square of the hatch, or where cargo is to be regularly discharged by grabs. A minimum thickness is given in the Rules when fork lift trucks are to be used.

In passenger ships the inner bottom plating is to be continued out to the ship's side in such a manner as to protect the bottom to the turn of the bilge (SOLAS requirement). Drainage is effected by means of wells situated in the wings, having a capacity not less than 0.17 cubic metres and extending to not nearer the shell than 460 mm.

Transverse framing – requirements

Plate floors are to be fitted at every frame in the engine room, under boilers, under bulkheads and toes of brackets to deep tank stiffeners, in way of change of depth in the double bottom and for the forward $0.25L$ (see Pounding). Elsewhere the spacing of plate floors is not to exceed 3 m with bracket floors at the remaining frames.

Side girders are to be fitted between the centre girder and margin plate extending as far forward and aft as is practicable. If the beam is more than 10 m but not more than 20 m, one side girder; for a beam over 20 m, two side girders. Additional side girders are provided in the engine room and pounding region.

The un··· ···an of the frames in bracket floors is not to exceed 2.5 m. Breadth ··· ···he frames to the centre girder and margin plate is to be 75% of ··· ···e girder. The brackets are to be flanged on their unsupported

··· requirements

···d at every frame under the the main engines and foremost ···ate frames outboard of the engine seating, also under boiler ···f brackets to deep tank stiffeners. Elsewhere, spacing of ··· 3.8 m except in the pounding region where they are on

DOUBLE BOTTOM CONSTRUCTION

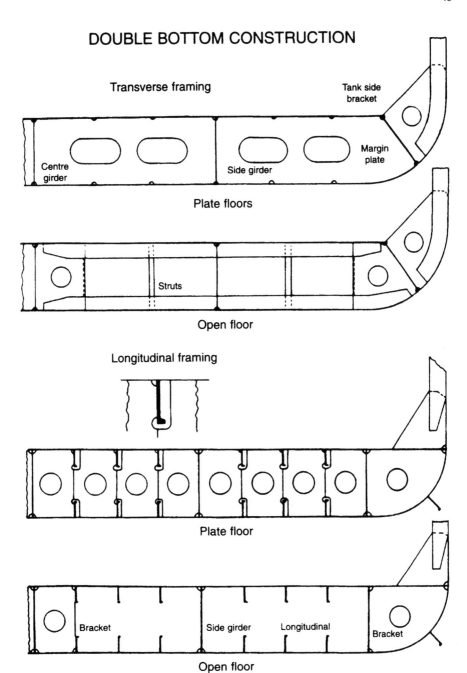

Transverse framing

Tank side bracket

Centre girder

Side girder

Margin plate

Plate floors

Struts

Open floor

Longitudinal framing

Plate floor

Bracket

Side girder

Longitudinal

Bracket

Open floor

alternate frames and where 'Heavy Cargoes' is assigned when maximum spacing is to be 2.5 m.

Between plate floors transverse brackets are to be fitted extending from the centre girder and margin plate to the adjacent longitudinal. Brackets are to be fitted at each frame at the margin plate and not more than 1.25 m apart at centre girder.

Side girders are to be fitted between the centre girder and margin plate extending as far forward and aft as practicable. For a beam more than 14 m but not more than 21 m, one side girder; for a beam over 21 m, two side girders. Additional side girders are provided in the engine room and pounding region. When the notation 'Heavy Cargoes' applies, spacing of side girders is not to exceed 3.7 m. Where L exceeds 215 m the bottom longitudinals should be continuous through the transverse bulkheads.

General

Sufficient holes are to be cut in the inner bottom non-watertight/non-oiltight floors and side girders to provide adequate ventilation and access. Their size should not exceed 50% depth of the double bottom and they should be circular or eliptical in shape.

Testing

Each compartment is to be tested with a head of water representing the maximum pressure which could be experienced in service or alternatively air pressure testing may be used.

Duct keel

A duct keel, consisting of 'twin' centre girders, is often fitted forward of the engine room, pipe lines being led through it.

The sides are not to be more than 2 m apart and and the inner bottom and keel plate are to be suitably stiffened to maintain continuity of strength transversely.

Bilge keels

Bilge keels are fitted at the turn of the bilge to help damp the rolling motion of the ship. They extend over a portion of the midship length of the ship and are positioned to minimize drag.

They are attached to a continuous flat bar, rather than directly to the shell, and their ends are gradually tapered and end on a frame or other shell-stiffening member.

DUCT KEEL

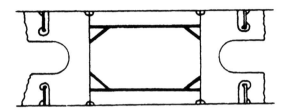

Insulation of tank top

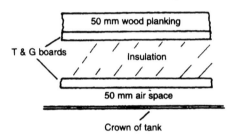

50 mm wood planking

T & G boards

Insulation

50 mm air space

Crown of tank

BILGE KEEL

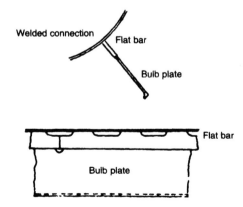

Welded connection

Flat bar

Bulb plate

Flat bar

Bulb plate

Pounding

Heavy pitching assisted by heaving as the whole ship is lifted in a seaway may subject the forepart to severe impact from the sea. The greatest effect is experienced in the light ship condition. To compensate for this the bottom over 30% forward is additionally strengthened in ships exceeding 65 m in length and in which the the minimum draught forward is less than $0.045L$ in any operating condition.

Bottom framed – longitudinally

Where the minimum draught forward is less than $0.04L$ in any operating condition, plate floors are to be fitted at alternate frames and side girders fitted at a maximum spacing of three times the floor spacing. If the minimum draught forward is between $0.04L$ and $0.045L$, plate foors are to be fitted at every third frame and side girders fitted at a maximum spacing of four times the floor spacing.

Bottom framed – transversely

Plate floors are to be fitted at every frame and side girders are to be fitted at a maximum spacing of three times the floor spacing. Half height side girders are to be provided midway between the full height side girders.

Pounding

Heaving

Pitching

Slamming ;

Strengthening of bottom forward

Plate floors at alternate frames

Girders
2.1 m

Plate floors at every frame

1.1 m — 1.1 m

Full height and half height girders

Panting

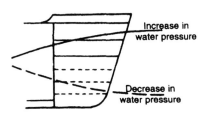

Increase in water pressure

Decrease in water pressure

Deck and shell plating

The deck and shell plating forms the watertight skin of the ship and is a major contributor to the longitudinal strength of the hull girder.

The plates are arranged in fore and aft lines around the hull, called 'strakes' which, for identification, may be lettered starting with the strake adjacent to the keel, this strake being A. The separate plates in the strakes may be numbered, usually from aft, thus 'C 12 port' will be the 12th plate from aft in the 3rd strake up from the keel on the port side.

The thickness of the plating depends, in general, on the length of ship and frame spacing. The midship thickness is to be maintained for $0.4L$ amidships and tapers gradually to an end thickness at $0.075L$ from the ends at the deck and at the ends for the shell. Special attention to thickness is required when decks are to carry excess loads, and to structural details in way of openings especially hatch corners. Abrupt changes of shape or section and sharp corners are to be avoided. Where plated decks are sheathed with wood or an approved composition, reductions in plate thickness may be allowed.

The upper edge of the sheerstrake is to be dressed smooth and kept free of isolated welded fittings or connections. Where the sheerstrake is rounded the radius is not to be less than 15 times the thickness of the plate.

The width of keel plate is to be $70B$ mm but need not exceed 1800 mm or be less than 750 mm. Its thickness is 2 mm greater than the adjacent bottom plating.

All openings in the shell and deck plating are to have well-rounded corners. Shell plating in way of hawse pipes is to be increased in thickness and the thickness of plates connected to the sternframe or propeller bracket are to be at least 50% greater than the adjacent plating.

At the ends of a ship, particularly at the bow, the width of strakes decreases and it is often desirable to merge two strakes into one, this being done by a 'stealer' plate. Also, at the ends of the ship where the keel plate terminates at the stem and sternframe these plates have been referred to as the 'shoe' and 'coffin' plate respectively.

If a ship is to be assigned a special features notation for navigation in 'first-year ice' (see page 6) the additional strengthening required includes an increase in plate thickness and frame scantlings in the waterline region and the hull bottom forward.

49

Shell expansion plan

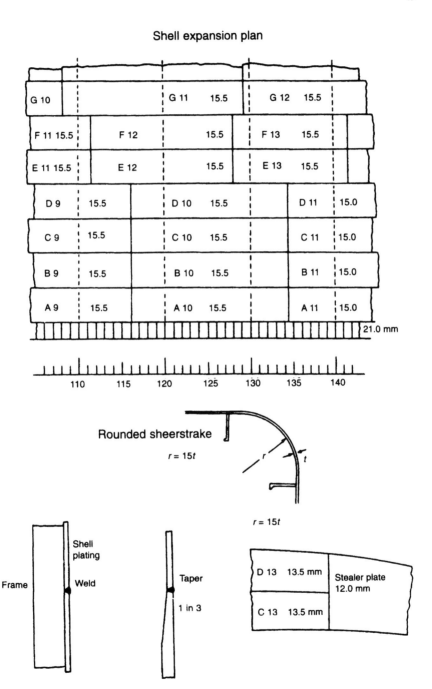

Rounded sheerstrake

$r = 15t$

$r = 15t$

Frame

Shell plating

Weld

Taper

1 in 3

D 13   13.5 mm
C 13   13.5 mm

Stealer plate 12.0 mm

Frames, beams and longitudinals

These are usually of offset bulb or inverted angle section. Longitudinal or transverse framing may be used except for the strength deck and bottom shell of ships exceeding 120 m in length, where longitudinal framing should be used.

The scantlings of transverse frames increase with depth and spacing. For identification purposes transverse frames may be numbered, usually from aft to forward and commencing at the transom floor. Aft of the transom floor they are usually lettered.

Where longitudinal framing is adopted the spacing and location determines the section modulus of the frame. At the side shell the lower longitudinal framing will have greater scantlings than that in the vicinity of the deck. Outside the peaks, side longitudinals are supported by webs spaced at not more than 3.8 m apart in ships of 100 m length or less, with increasing spacing permitted for longer ships. Deck longitudinals are similarly supported by transverses. Where L exceeds 215 m the bottom and deck longitudinals should be continuous through transverse bulkheads, with the longitudinals attached to the bulkhead in such a manner as to maintain direct continuity of longitudinal strength (see Tankers, page 75).

Where the deck is transversely framed the deck beams are to be fitted at every frame. Deck beams are required to support the deck and any loads it carries and to act as struts assisting in holding the sides of the ship apart against the inward pressure of the sea.

Beam knees are fitted to provide an efficient connection between the side frames and deck beams. They provide a small amount of resistance against racking stresses. The size of the knees is determined by the scantlings of the frame and beam and are given in the Rules.

Tankside bracket

The lower end of the side frame is to be connected to the tank top or margin plate by a bracket as illustrated. The Rules detail the required thickness, length of overlaps, size of flange or edge stiffener etc.

Deck girders

Deck beams are supported by longitudinal deck girders which are usually built sections. The built section girders are supported by 'tripping brackets' at every second beam if of unsymmetrical section, and at every fourth beam if of symmetrical section.

Within the forward 7.5% of the ship's length these deck girders are more closely spaced on the weather and forecastle decks at 3.7 m. Elsewhere the spacing is arranged to suit the deck loads carried, deck openings and pillar arrangements.

In way of hatches fore and aft side girders are fitted to support the inboard ends of the deck beams. At the ends of the hatches heavy deck beams are connected at the intersection of the hatch side girder by horizontal gusset plates (see illustration page 79).

Beam knees

Tankside
brackets

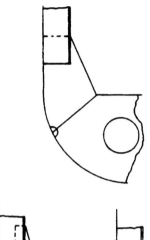

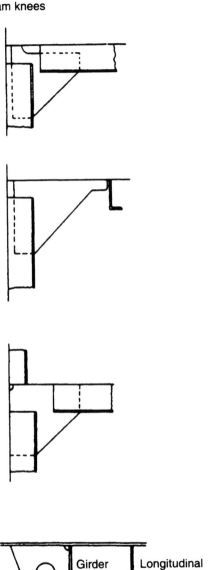

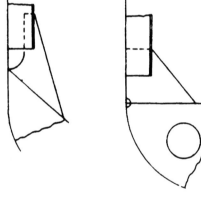

Girder   Longitudinal

Tripping bracket at alternate
beams or equivalent
(long. framing)

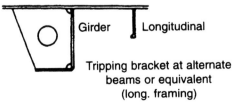

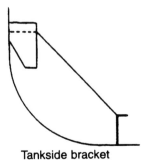

Tankside bracket
tanker

Panting

This is a stress which occurs at the ends of a ship due to variations in water pressure on the shell plating as the ship pitches in a seaway. The effect is accentuated at the bow when making headway.

Additional strengthening is provided in the forepeak structure, the transverse side framing being supported by any, or a combination, of the following arrangements:

(i)   side stringers spaced vertically about 2 m apart and supported by 'panting beams' fitted at alternate frames. These 'panting beams' are connected to the frames by brackets and, if long, supported by a partial wash bulkhead at the centreline; or
(ii) side stringers spaced vertically about 2 m apart and supported by web frames; or
(iii) perforated flats spaced not more than 2.5 m apart. The area of perforations being not more than 10% of the total area of the flat.

Abaft the forepeak, panting stringers are fitted in line with each panting stringer or perforated flat in the forepeak extending back over a distance of 0.15L from forward. These stringers may be omitted if the side shell plating thickness is increased by 15% for ships of 150 m in length or less, or 5% for ships of 215 m length or more, with intermediate length reductions being determined by interpolation. However, if the unsupported length of side frames is greater than 9 m panting stringers must be fitted in line with alternate stringers or flats in the forepeak over 0.2L from forward, whether the shell plating is increased or not.

In the aft peak space, similar panting arrangements are required but the vertical spacing of stringers may be up to 2.5 m apart.

ARRANGEMENT FORWARD OF COLLISION BULKHEAD

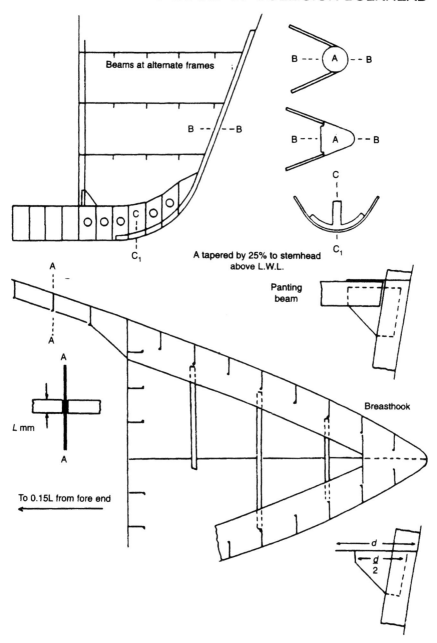

Beams at alternate frames

B - - - B

A tapered by 25% to stemhead
above L.W.L.

Panting
beam

Breasthook

L mm

To 0.15L from fore end

Stems

The lower portion of the stem may be formed by a solid round bar to which the side shell plates are welded. From the waterline area upward the stem is formed by radiused plates stiffened between decks by short horizontal webs known as 'breast hooks'. Where the radius is large, further stiffening may be provided by a vertical stiffener. The thickness of the plate at the waterline will be heavier than that of the adjacent shell but tapers to that of the side shell at the stem head.

Bulbous bows

Constructional arrangements are dependent upon the shape of the bow. In general the protrusion forms a continuation of the side shell. Plate floors are fitted at every frame and transverse webs at about every fifth frame in long bulbs. A centreline web is also fitted and in larger bulbs this becomes a full wash bulkhead. In all bulbous bows horizontal diaphragm plates are fitted not more than 1 m apart.

PANTING – Alternative arrangements

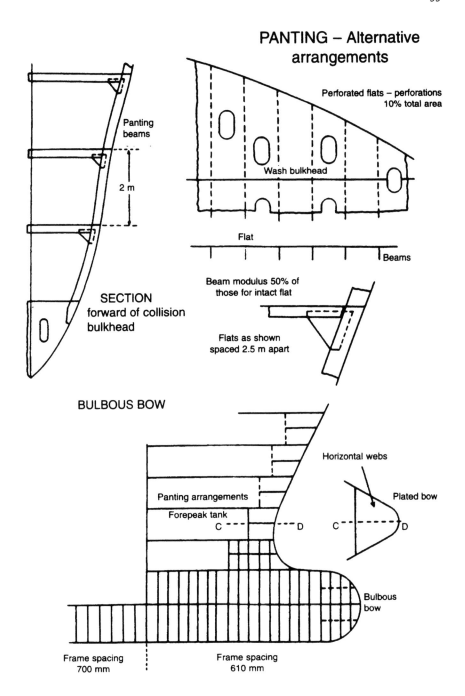

Panting beams

2 m

SECTION forward of collision bulkhead

Perforated flats – perforations 10% total area

Wash bulkhead

Flat

Beams

Beam modulus 50% of those for intact flat

Flats as shown spaced 2.5 m apart

BULBOUS BOW

Horizontal webs

Plated bow

Panting arrangements

Forepeak tank

C   D   C   D

Bulbous bow

Frame spacing 700 mm

Frame spacing 610 mm

Hawse pipes

Hawse pipes and anchor pockets are to be of ample thickness and of suitable size and form to house the anchors efficiently and preventing, as much as practicable, slackening of the cable or movements of the anchor being caused by wave action. The shell plating and framing in the way of hawse pipes often requires reinforcement. Substantial chafing lips are required at the ends of the hawse pipe at the deck and shell. These are often steel castings welded to the ends of the tubular steel hawse pipe and shell. Alternatively the hawse pipe may be an integral cast steel structure.

Bow thrusters

Directional control at low speeds is a highly desirable feature for many ships. In particular for the berthing of large ships and the accurate positioning of research ships and work platforms. This directional control may be obtained by the use of bow thrusters.

These units may consist of:

(a)   A shrouded propeller, where the shroud is movable and directs the thrust in a desired direction.
(b)   A transverse tunnel or duct through the ship near the bow in the narrow forward section. A reversible or controllable pitch impeller is fitted on the ship's centreline within the tunnel and this acts as a pump discharging large quantities of water to either side and creating the desired athwartships thrust.

In some ships a similar bow thruster is provided aft near the stern.

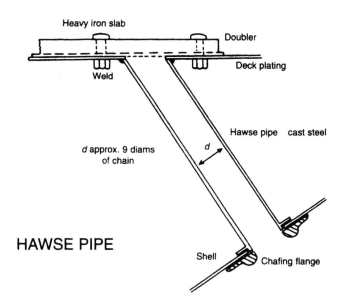

Heavy iron slab

Doubler

Weld

Deck plating

Hawse pipe   cast steel

d approx. 9 diams
of chain

d

HAWSE PIPE

Shell

Chafing flange

BOW THRUSTER UNIT

Grating
over
opening

Propeller

Collision
bulkhead −

Aft end structure

The illustration shows the general arrangement aft for a ship with a cruiser-type stern and aft engine room. The floor to which the rudder post (which is carried up into the main hull) is fitted is heavier with more substantial stiffening than the adjacent floors. This floor has been referred to as the 'transom floor'.

Abaft the transom floor a heavy centreline girder and side girders are fitted. Transverse plate floors are fitted at every frame abaft the aft peak bulkhead and carried up to the steering flat. Panting arrangements within the aft peak are as detailed on page 52.

GENERAL ARRANGEMENT AFT

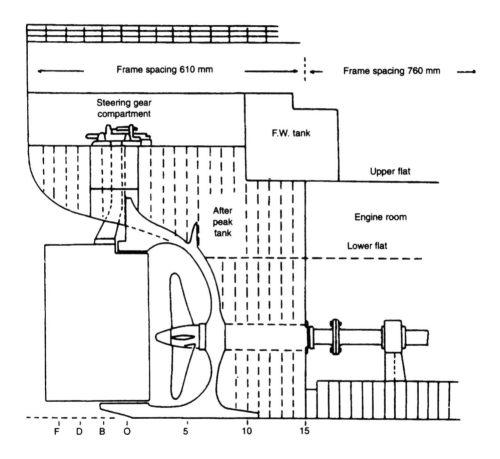

Stern frames

Stern frames may be cast, forged or fabricated from steel plate and sections. Forged stern frames are not generally found on larger ships and for these vessels a cast stern frame may need to be cast in more than one piece. The castings may be welded together when erected at the shipyard. Thermit welding described on page 32 is used for this purpose.

The use of a welded connection is illustrated in the figure showing a cast stern frame and semi or balanced rudder.

All stern frames are to be efficiently attached to the adjoining hull structure. In addition to the rudder post being attached to the transom floor the propeller post is also carried up into the hull and attached to a floor, and the lower part of the stern frame is extended forward to provide an efficient connection to the flat plate keel. If the attachment is not substantial, the propeller supported by the stern frame may set up serious vibrations in this area.

CAST STEEL STERN FRAMES

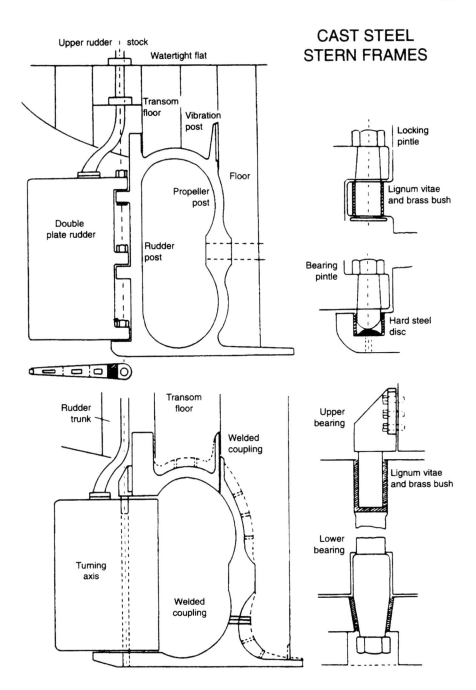

Upper rudder stock

Watertight flat

Transom floor

Vibration post

Floor

Propeller post

Double plate rudder

Rudder post

Locking pintle

Lignum vitae and brass bush

Bearing pintle

Hard steel disc

Rudder trunk

Transom floor

Welded coupling

Upper bearing

Lignum vitae and brass bush

Turning axis

Welded coupling

Lower bearing

Fabricated stern frame

Many larger stern frames are fabricated with the rudder and propeller posts built into the adjacent hull structure. A fabricated propeller post arrangement is illustrated with accompanying cross-sections in way of various frames.

Note that where a balanced rudder is fitted the rudder post is omitted and the unsupported sole piece is then required to be of a more substantial section.

FABRICATED STERN FRAME

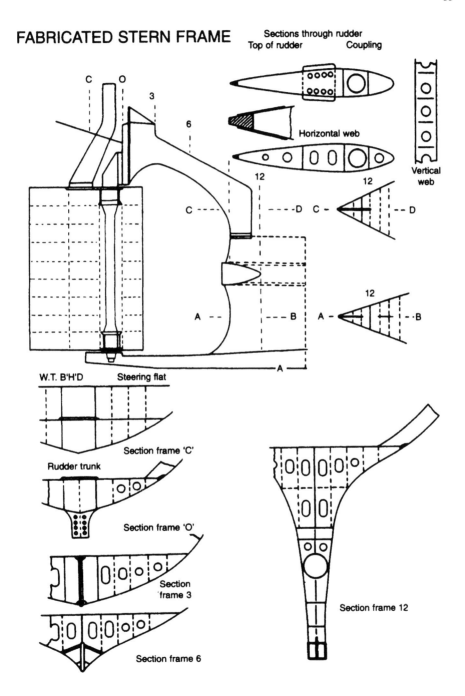

Sections through rudder
Top of rudder   Coupling

Horizontal web

Vertical web

W.T. B'H'D   Steering flat

Section frame 'C'

Rudder trunk

Section frame 'O'

Section frame 3

Section frame 6

Section frame 12

Stern frame for twin screw ship

The stern frame of a twin or quadruple screw ship does not have to support the tail-shaft and propeller, its only function is to support the rudder.

The various illustrations show a cast stern frame and a fabricated stern frame for a twin screw ship. In the latter case, sections in way of the stern frame at various frames forward and abaft of the transom floor have been illustrated.

In a twin screw ship the tail shafts may be enclosed by bossings, and supported at the ends by a spectacle frame, or the shafts may be exposed after leaving the hull, with the after ends supported by an 'A' bracket or frame. Occasionally both may be found in the same ship, a short portion of the shafts being enclosed by a bossing and the remainder of the shafts exposed and supported by an 'A' frame.

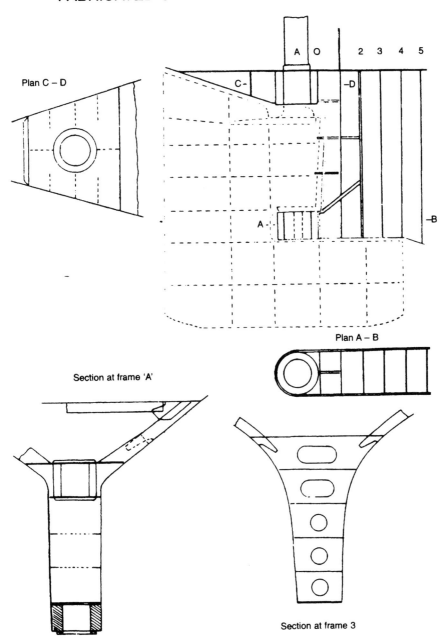

FABRICATED STERN FRAME TWIN SCREW

Plan C – D

Plan A – B

Section at frame 'A'

Section at frame 3

Spectacle frame

A 'spectacle frame' may consist of two castings attached to the stern section and welded together at the ship's centreline, or, in a very large ship, extend only far enough inboard of the shell plating for an adequate connection to be obtained with the adjacent structure. The illustrations show the initial curvature of the shell plating at the after end to that of the actual 'spectacle frame'. It will be seen that the bossing is continuous with the shell plating and is faired to a fine trailing edge abaft the spectacle frame so as to allow the flow of water to the propeller to be as undisturbed as possible.

'A' brackets

Where plated bossings or a spectacle frame are not used an 'A' bracket is fitted. The two streamlined struts forming the bracket will usually pass through the shell and the inboard connection is then made to a system of brackets and frames so that stresses are transmitted to the adjacent structure. Watertightness is maintained by welding round the strut where it passes through the hull. To ensure that the 'A' bracket is rigid there is an angle of 60° to 90° between the struts.

Both 'spectacle frames' and 'A' brackets are used but the former is more common in large twin screw ships.

PROPELLER FRAMES
TWIN SCREW VESSELS

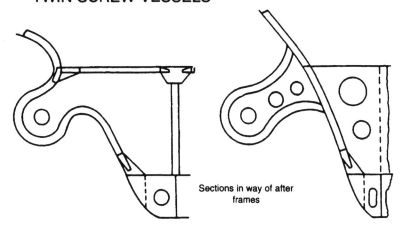

Sections in way of after
frames

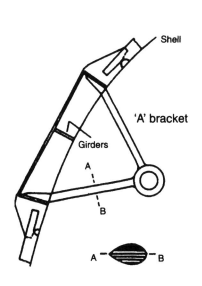

Shell

'A' bracket

Girders

A

B

A —◯— B

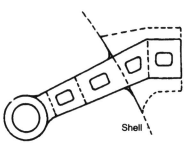

Shell

Cast spectacle frame

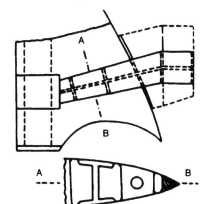

A

B

A - - - - - - - B

Rudders

There are several illustrations of rudders in association with stern frames which show various rudder types, their construction and bearings.

The shape and size of the rudder plays an important part in its efficiency. The area may be of the order of 2% of the product of ship's length and designed draught. Most rudders are semi-balanced having up to 20% of their area forward of the turning axis, but balanced rudders with 25 to 30% of their area forward of the turning axis and unbalanced rudders with all area aft of the turning axis are also fitted. Since the vertical dimensions of a rudder are necessarily restricted, the fore and aft dimension must be increased to obtain the desired area and the distance to the centre of lateral resistance from the turning axis is increased. The torque needed to turn the rudder is thus greater and the object of balance is to reduce this by providing area forward of the turning axis which reduces the distance to the centre of lateral resistance.

On small ships a single plate rudder may be used but larger ships have faired double plate rudders. The construction of a faired double plate rudder may have a cast frame but more often consists of a fabricated frame of vertical and horizontal plate webs with a solid or tubular mainpiece which coincides with the turning axis. The faired side plates are welded to the frame and top and bottom plates fitted to provide a watertight and buoyant structure. The rudder stock is usually of solid round or tubular section ending in a flanged coupling which can be bolted to a matching flange on the top of the rudder.

A simple unbalanced rudder turns on 'pintles' which are fitted in 'gudgeons' attached to the rudder post. The top pintle being a 'locking pintle' which prevents vertical movement of the rudder and the bottom pintle being a 'bearing pintle' which carries the weight of the rudder.

Many types of rudder can be met with in practice, a number of which may be patent types.

Steering gear

All ships are to be provided with two independent means of moving the rudder. With power operated steering gears the time taken to put the rudder from 35° on one side to 35° on the other side, or 35° on either side to 30° on the other side, is not to exceed 28 sec at the maximum service speed.

Auxiliary steering gear is to be of adequate strength, sufficient to steer the ship at a navigable speed and capable of being brought speedily into operation in an emergency.

The classification society rules cover all aspects of rudder construction i.e. stock size, rudder scantlings, pintle dimensions etc. Both the rudder and stern frame scantlings are increased if the ship is to operate in ice. SOLAS requirements dictate steering arrangements including emergency provisions.

BALANCED RUDDER TWIN SCREW

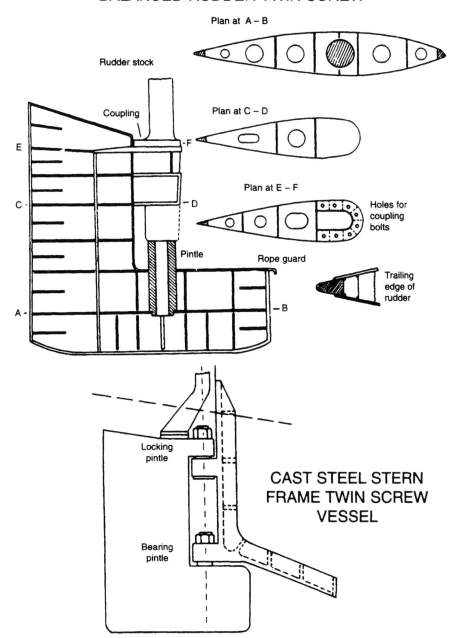

Plan at A – B

Rudder stock

Coupling

Plan at C – D

E

Plan at E – F

C · D

Holes for coupling bolts

Pintle

Rope guard

Trailing edge of rudder

A · B

Locking pintle

Bearing pintle

CAST STEEL STERN FRAME TWIN SCREW VESSEL

Stern tubes

Stern tubes are fitted to provide a bearing for the 'tail end shaft' and to enable a water-tight gland to be fitted at an accessible position.

Two types of stern tube are in common use, one with water-lubricated bearings having the aft end of the tube open to the sea, and the other with oil-lubricated metal bearings. The latter is often preferred when the machinery is aft and the short shaft is to be relatively stiff and only small deflections are tolerated. It is also used when the ship frequently operates in water containing sand or sediment and the wear down on the non-metal bearing material could be excessive.

The tube is usually constructed of cast steel with a flange at its forward end and a thread at the aft end. The tube is inserted from the forward end and the flange is bolted to the aft peak bulkhead with a gasket to ensure watertightness. A large nut is placed on the thread at the after end, tightened and secured to the propeller post.

In a water-lubricated stern tube the forward end is fitted with a watertight gland, and the after end is fitted with a brass bush into which strips of bearing material are set. Traditionally the bearing material used was a very hard wood 'lignum vitae', but laminated plastics are now often used. A bronze liner is fitted to the tail end shaft, usually for the full length of the stern tube.

An oil-lubricated stern tube has glands at both ends, the after gland being self adjusting since it is not accessible in service. The bearings are of white metal and a bronze liner is therefore unnecessary on the tail end shaft. In the type illustrated a flange is attached to the propeller so that it rotates with the shaft and oiltightness is obtained by a rotating gland.

The after end of the tail end shaft is tapered to receive the propeller boss and a key is provided to transfer the torque from the shaft to the propeller. A nut, fitted with a locking plate, secures the propeller in position and, as an additional safeguard, it is fitted with a left-hand thread in association with the right-hand propeller, or a right-handed thread in association with a left handed propeller.

To remove the propeller and the tail end shaft, the propeller should be slung, after removing the rope guards, and the propeller nut slacked back. The propeller is then started from the shaft by driving steel wedges between the boss and the propeller post. When it is free the propeller nut is removed. The intermediate shaft (length of shaft next to the tail end shaft) is removed. The tail end shaft may then be withdrawn into the tunnel and the propeller can then be removed from the aperture.

Withdrawal of the tail end shaft is generally required every five years.

TAIL END SHAFT AND STERN TUBE

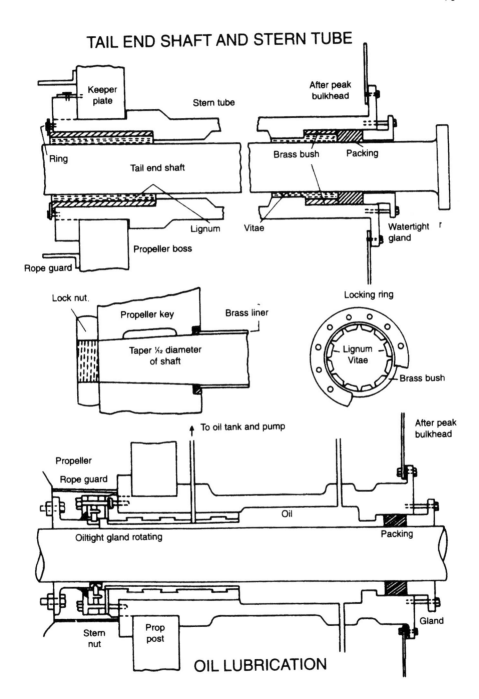

OIL LUBRICATION

Watertight bulkheads

Transverse watertight bulkheads which divide a ship into a number of watertight compartments are of great importance for the following reasons:

(1) Strength: they give large structural support, resist any tendency to deformation (racking) and assist in spreading the hull stresses over a large area.
(2) Fire: confines conflagration to particular regions.
(3) Subdivision: divides a ship into a number of watertight compartments.

All ships are to have a collision bulkhead, situated not less than $0.05L$ nor more than $0.08L$ for cargo ships ($0.05L + 3$ m for passenger ships) from the fore end of the load waterline, an after peak bulkhead enclosing the stern tubes in a watertight compartment and a bulkhead at each end of the machinery space.

Additional watertight bulkheads are to be fitted in cargo ships depending on the length of the ship. Structural compensation is to be made where the number of bulkheads is below Rule number. The spacing of additional bulkheads in passenger ships is dictated by the damage subdivision requirements of SOLAS. Many bulk carriers which have reduced freeboards assigned under the International Load Line Convention have hold bulkhead spacing which is dictated by subdivision requirements.

All watertight bulkheads are to extend to the uppermost continuous deck, except for the after peak bulkhead which may terminate at the first deck above the load waterline provided this deck is made watertight to the stern or to a watertight transom floor. However, where the freeboard is measured from the second deck, watertight bulkheads need only be taken up to that deck, except the collision bulkhead which is maintained to the uppermost continuous deck.

The thickness of bulkhead plating is greater at the bottom than the top as it depends on the head of water that the horizontal strakes of plating would be subject to if the compartment was flooded. It also depends on the spacing of the vertical stiffeners whose sectional modulus increases with unsupported length of span and depth of bulkhead. The scantlings of the fore peak bulkhead are greater than those of the other watertight bulkheads.

TRANSVERSE WATERTIGHT BULKHEAD

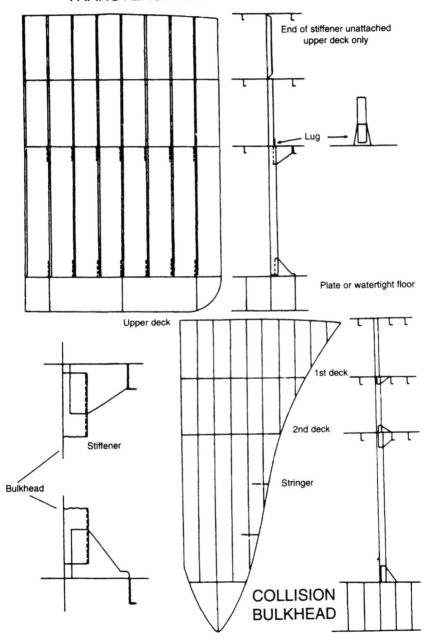

End of stiffener unattached
upper deck only

Lug →

Plate or watertight floor

Upper deck

1st deck

2nd deck

Stiffener

Stringer

Bulkhead

COLLISION BULKHEAD

If there are openings in watertight bulkheads, watertight doors with suitable framing must be fitted, and additional stiffening in way of the doors must be fitted so that strength is the same as that of the unpierced bulkhead.

Pipes and valves attached directly to the bulkhead plating are to be secured by studs screwed through the plating or by welding.

Corrugated plating is frequently used for bulkheads, particularly in tankers, bulk carriers and tank spaces of other ships. The corrugations are usually trapezoidal in shape and such bulkheads afford a considerable saving in welding, are less susceptible to corrosion and provide easier tank cleaning.

Superstructure bulkheads are occasionally swedged, the spacing of the vertical swedges being similar to that of stiffeners.

Testing

Watertight bulkheads including recesses and flats are to be hose-tested on completion. Peak bulkheads not forming boundaries of tanks are to be tested by filling peaks with water to the level of the load waterline.

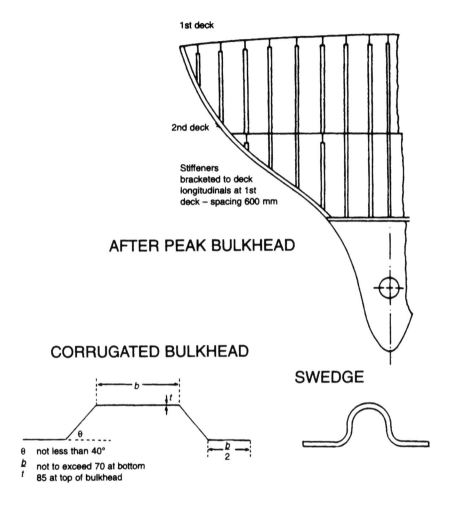

1st deck

2nd deck

Stiffeners
bracketed to deck
longitudinals at 1st
deck – spacing 600 mm

AFTER PEAK BULKHEAD

CORRUGATED BULKHEAD

b

t

θ

θ   not less than 40°
b   not to exceed 70 at bottom
t   85 at top of bulkhead

b / 2

SWEDGE

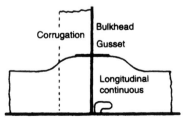

Corrugation

Bulkhead

Gusset

Longitudinal
continuous

Arrangement of bottom longitudinal in way of
O.T. transverse bulkhead

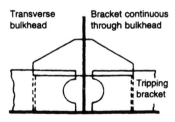

Transverse
bulkhead

Bracket continuous
through bulkhead

Tripping
bracket

Longitudinal with ends scalloped

Pillars, girders and non-watertight bulkheads

The primary function of pillars and girders is to transmit the deck loads to the bottom structure where the distributed loads are supported by the upthrust of the water pressure. They also tie the ship together vertically, thus preventing the flexing of the decks in response to the bending of the side frames under varying water pressures and vertical accelerations in heavy weather. Normally a pillar will be in compression although in certain cases it is possible for the pillar to be subjected to tension and even side loads resulting from movement of cargo when rolling.

In cargo holds large widely-spaced tubular pillars are usual. These tend to reduce 'broken stowage' and such an arrangement may be referred to as 'massed pillaring'.

Pillars are to be fitted in the same vertical line wherever possible and effective arrangements made to distribute the load at the head and heel. Pillars are to be securely bracketed at their heads with doubling plates fitted at the head and heel, which should be over the intersection of a plate floor and side girder not having manholes under the pillar. Where pillars are not arranged directly above the intersection of floors and girders then partial floors and short intercostal girders are to be fitted.

Pillars are to be fitted below deck houses, windlasses, winches etc. to give the necessary support.

In lieu of pillars non-watertight pillar bulkheads may be fitted on the ship's centre-line. They usually extend from the transverse bulkhead to the hatch coaming – such an arrangement facilitates the fitting of shifting boards when carrying grain cargoes.

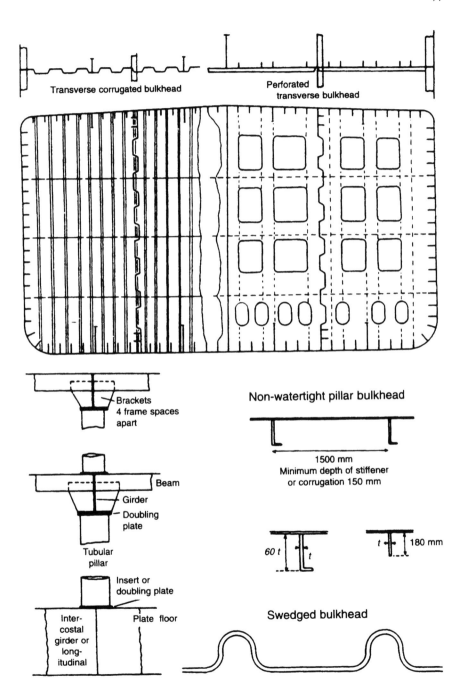

Transverse corrugated bulkhead

Perforated
transverse bulkhead

Brackets
4 frame spaces
apart

Beam

Girder

Doubling
plate

Tubular
pillar

Insert or
doubling plate

Intercostal
girder or
longitudinal

Plate floor

Non-watertight pillar bulkhead

1500 mm
Minimum depth of stiffener
or corrugation 150 mm

$60\,t$ t

t 180 mm

Swedged bulkhead

Hatchways

Hatchways in the majority of dry cargo ships extend across the deck for approximately one-third of the beam. In special types of ship, e.g. container ships, colliers etc., much wider hatchways are fitted as will be seen from the separate cross-sections of these types.

Special arrangements must be made to compensate for the structural discontinuities caused by these large openings; insert plates of increased thickness may be required at the hatch corners as shown. The arrangement of hatch coaming and adjacent structure with rounded hatch corners is illustrated. Note that the hatch coaming should be extended beyond the corner to form a bracket. The deck opening corners should be well-rounded to a reasonable radius, or eliptical or parabolic in shape, to avoid a concentration of stress at these points, see page 23.

The deck plating forms an important structural member, especially at the strength deck, in resisting longitudinal stress, but only the plating clear of the hatches can be considered in this respect. If hatchways are made unduly wide, the effective width of this plating is reduced and the thickness has to be increased in order that the cross-sectional area of the steel is maintained.

In addition to the plating, beams will also be cut at hatchways and the ends of the half beams will be connected to the hatch coamings and supported by deck girders. The deck girders are usually integral with the hatch coamings as illustrated. At the ends of the hatchways, in the case of transverse framing and at the sides in the case of longitudinal framing deep hatch end beams will be fitted to support the coamings. A pillar will often be placed near the hatch corners at the intersection of the deck girder and a strong beam.

HATCH CORNERS

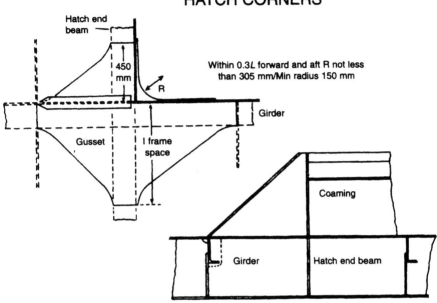

Within 0.3L forward and aft R not less than 305 mm/Min radius 150 mm

Hatch end beam

450 mm

R

Girder

Gusset

I frame space

Coaming

Girder

Hatch end beam

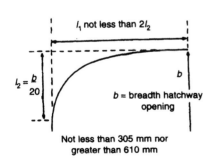

l_1 not less than $2l_2$

$l_2 = \dfrac{b}{20}$

b

b = breadth hatchway opening

Not less than 305 mm nor greater than 610 mm

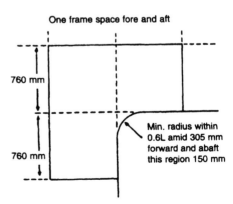

One frame space fore and aft

760 mm

760 mm

Min. radius within 0.6L amid 305 mm forward and abaft this region 150 mm

Hatch coamings

The International Load Line Convention requires that hatch coamings on weather decks are not to be less than 600 mm in height in Position 1 and 450 mm in height in Position 2.

Position 1   Hatchways on exposed freeboard decks, and exposed superstructure decks within forward 0.25*L*.
Position 2   Hatchways on exposed superstructure decks abaft 0.25*L*.

In practice the coamings may be higher to meet statutory requirements relating to dock safety, unless temporary fencing is provided when working cargo.

Hatch coamings of more than 600 mm height are to be stiffened on their upper edge by a horizontal stiffener not less than 180 mm in depth. Additional support is given by brackets or stays from the horizontal stiffener to the deck at intervals of not more than 3 m. Coamings less than 600 mm in height are to be stiffened on their upper edge by a substantial moulding. Side coamings are to extend to at least the lower edge of the deck beams and the lower edge is to be flanged to avoid damage to cargo and the fraying of any runner wires.

Hatch covers

Where hatch covers are used in conjunction with portable beams and indirect securing arrangements the wooden hatch covers will have a thickness of not less than 60 mm where the span is 1.5 m, and 82 mm with a span of 2 m, and proportionate thickness for intermediate spans. The ends of the hatch covers will be protected by galvanized steelbands.

The hatch covers rest on beams as illustrated. The hatch beams consist of webs stiffened at their upper and lower edges. The ends of the webs are to be doubled or inserts fitted. Those beams which carry the ends of the hatch cover (king beams) are to be fitted with a 50 mm vertical flat on the upper surface and the bearing surface for the covers is to be not less than 65 mm. The portable beams are supported at their ends by carriers or sockets having a minimum bearing surface of 75 mm.

Covering tarpaulins, at least two, are held in position by battens running along the side and ends of the hatches. These are secured by tapered toughwood wedges and cleats which are set to take the taper of the wedges. Steel bars or other locking devices are to be provided to secure each section of hatch covers after battening down. Hatch covers over 1.5m in length have two such securing devices.

Roller or sliding beams may be fitted with the coaming's horizontal stiffener forming the trackway. Also, portable stiffened plates may be used in lieu of wood hatch boards.

Portable hatch covers of the type described above incur a freeboard penalty under the International Load Line Convention and involve labour intensive handling and thus they are unlikely to be found on larger modern ships.

HATCHWAYS

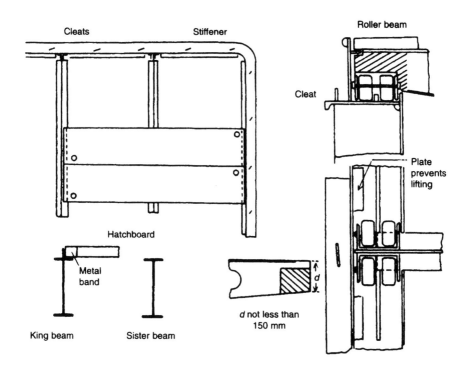

Cleats

Stiffener

Roller beam

Cleat

Hatchboard

Metal band

King beam

Sister beam

d not less than 150 mm

Plate prevents lifting

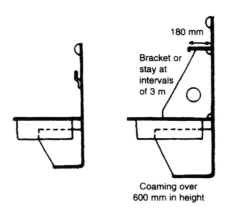

180 mm

Bracket or stay at intervals of 3 m

Coaming over 600 mm in height

Patent steel hatch covers

Patent steel hatch covers having direct securing arrangements, e.g. MacGregor Steel Hatches are a great improvement on the portable type previously described and are universally fitted for weather decks. They consist of plated covers stiffened by webs or stiffeners, watertightness being obtained by gaskets and clamping devices.

Securing cleats and cross joint wedges, together with suitable jointing material are to be fitted, the cleats are to be spaced to ensure weathertightness with a minimum of two per panel at the sides and with one arranged adjacent to each corner at the hatch ends. The cross joint wedges are spaced about 1.5 m apart.

The portable sections of folding covers are connected to one another and can easily and quickly be rolled into or out of position, leaving clear hatchways and decks. The normal practice is for the lengthwise opening of patent hatches but sideways opening hatches are found on some particular types of ships, e.g. OBO carriers, see pages 85 and 125. Patent steel hatch covers may be operated manually or hydraulically. The illustration shows a folding patent steel hatch cover.

The wheels at the sides of the hatch sections, eccentric rollers, are used for raising the hatch section clear of the coaming and for rolling it along the coaming trackway. As shown the axles of these wheels are so adjusted that when the hatch is in the closed position the weight is no longer borne by them. The jointing fits tightly on the coaming and the hatch is made completely weathertight by fitting and securing the cleats.

The roller is used when the hatch cover is pulled into its stowage position. It engages on the plate edge at the ends of the hatchway and the hatch section is turned into the vertical. Wires, chains or bars attached to the stub axles of these rollers at the centre of the wheel enable all the hatch sections to be drawn back and forth together.

The cross joints are made weathertight as shown with cross joint wedges.

Details of gas tightness with reference to OBO carriers are illustrated on page 85.

STEEL HATCHES

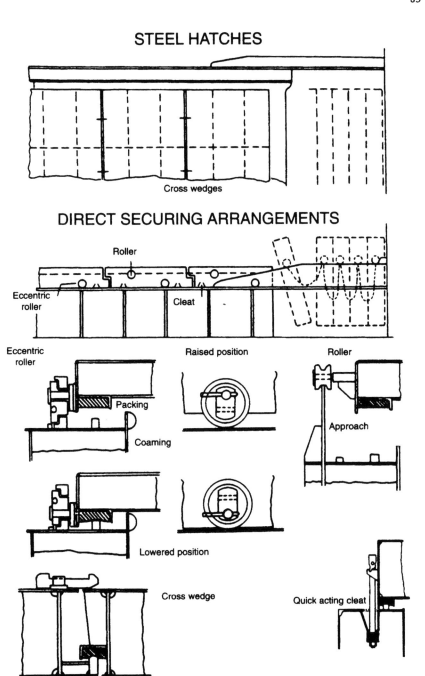

Cross wedges

DIRECT SECURING ARRANGEMENTS

Roller

Eccentric roller

Cleat

Eccentric roller

Raised position

Roller

Packing

Coaming

Approach

Lowered position

Cross wedge

Quick acting cleat

Peak, deep and topside tanks

The closing arrangements for deep tanks are illustrated. Access to peak and topside tanks is usually by a manhole which is closed with a steel cover bolted on a gasket to ensure it is watertight.

A centreline bulkhead is to be fitted in deep tanks which are designed for the carriage of liquid cargoes, or as oil fuel bunkers, and which extend the full width of the ship. This bulkhead may be solid or perforated. If intact the construction is similar to that of the tank's transverse bulkheads. If the bulkhead is perforated, the perforations are to be between 5 and 10% of the total bulkhead area, and the construction may be lighter.

Horizontal girders which are continuous on the bulkheads and ship's side may be fitted in deep tanks to support the vertical stiffeners.

Particular attention is to be given to the structural arrangements within topside tanks. A transverse is to be fitted within the tank in line with the end of the main cargo hatchways.

Deep tanks are to be tested by a head of water equal to the maximum to which the tank may be subjected, but not less than 2.45 m above the crown of the tank.

Manhole covers fitted outside a tank are circular or elliptical but when fitted inside the tank they are elliptical in order to facilitate withdrawal. Typical manholes have a 450 mm by 600 mm opening.

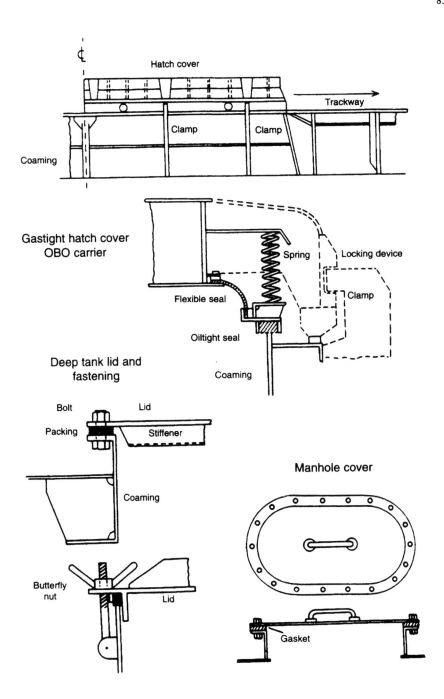

Hatch cover

Trackway

Clamp   Clamp

Coaming

Gastight hatch cover
OBO carrier

Spring   Locking device

Flexible seal   Clamp

Oiltight seal

Coaming

Deep tank lid and
fastening

Bolt   Lid

Packing   Stiffener

Coaming

Manhole cover

Butterfly
nut   Lid

Gasket

Superstructures

The International Load Line Convention requires that sea-going ships are fitted with a forecastle or have increased sheer forward to meet a required minimum bow height.

In practice most ships are fitted with a forecastle which extends at least $0.07L$ abaft the forward perpendicular.

The thickness and other scantlings of the fore ends and sides of superstructures are heavier than the aft ends.

As mentioned on page 22, particular attention is given to the ends of superstructures particularly where the superstructure exceeds $0.15L$ and is within 50% of the ship's length amidships. At the ends of such a bridge structure the upper deck sheerstrake is increased by 20%, the upper deck stringer plate by 25%, and the bridge side plating by 25%. The latter plating is tapered into the sheerstrake plating with a generous radius. The tapered plating is stiffened along its upper edge and supported by plate webs not more than 1.5m apart. At the ends of a shorter bridge superstructure local stresses can still be high and therefore the upper deck sheerstrake is still increased by 20% and the upper stringer plate by 20%. If the poop front comes within 50% of the ship's length amidships stiffening is as for a long bridge.

Where deckhouses are fitted above superstructures, webs or short transverse bulkheads are fitted at the superstructure sides to support them. They may also be fitted in way of large openings, boat davits and other points of high loading.

Portholes, side lights or side scuttles

When fitted below the freeboard deck they are provided with a steel deadlight which fastens onto a rubber gasket to ensure watertightness in the event of the glass being broken. They must be fitted a minimum of $0.025B$ mm or 500 mm, whichever is greater, above the deepest load line.

SUPERSTRUCTURE BULKHEAD

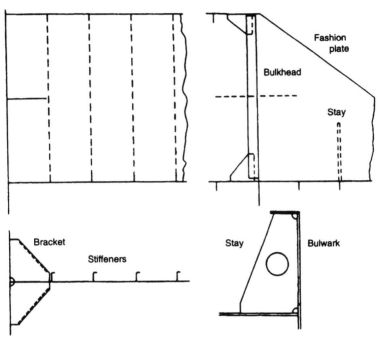

Fashion plate

Bulkhead

Stay

Bracket

Stiffeners

Stay

Bulwark

PORTHOLE

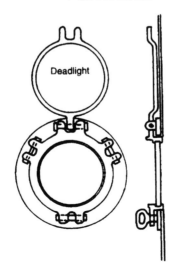

Deadlight

Watertight doors

Both the hinged and vertical sliding type are shown. Hinged watertight doors are not permitted below the waterline. Sliding doors may be hand-operated, but in most modern ships they are hydraulically controlled. SOLAS requirements stipulate that they be capable of being operated when the ship is listed 15°, and be opened and closed from a position above the bulkhead deck or locally. They are also to have an indicator at the above deck operating position showing if they are open or closed.

Watertight doors for passenger ships are tested to a head of water equivalent to their depth below the bulkhead deck before they are installed in the ship. They and other watertight doors are hose-tested in place.

Watertight tunnels

Unless the ship's machinery is right aft a watertight tunnel will be fitted to enclose the propeller shaft. The tunnel protects the shaft from the cargo and is large enough to permit access for examination and servicing of the shaft (see illustration on page 99).

A sliding watertight door will be fitted at the forward end leading to the engine room. Also, at the aft end a trunk leading to the bulkhead deck is often fitted since there must be two means of escape from the shaft tunnel.

The thickness of the tunnel plating and stiffener scantlings are determined in a similar manner to that for watertight bulkheads.

The tunnel top is lighter if it is rounded rather than flat, but must be increased in thickness under hatchways unless it is sheathed with timber.

WATERTIGHT DOORS

Hinged door

Dogs

Wedges

Sliding door

Gunmetal nut

Frame

Cargo doors

Cargo doors are fitted in certain trades to provide access to tween deck spaces, e.g. direct loading by fork lift truck from the quay into the tween deck.

Openings are cut in the side shell plating and arrangements must be made to maintain the strength, particularly in a longitudinal direction. The corners of all openings are to be well-rounded to avoid stress concentrations.

Illustrated are:

(1)   A cargo port, manually operated, secured by closely spaced dogs or bolts. This arrangement is typical of the type fitted to facilitate the loading of stores etc.
(2)   A patent hydraulically operated sliding door shown in the open and closed positions. This type is simple and fast to operate and is self closing since the door is forced against the perimeter of the opening door to the eccentric path of its guide rollers.
(3)   A swing door. This type of door may be fitted at the sides of the ship to give access to the tween deck or at the stern to give access for vehicles e.g. Ro-Ro ships. In the latter case the ramps will be a separate item of equipment.

CARGO DOORS

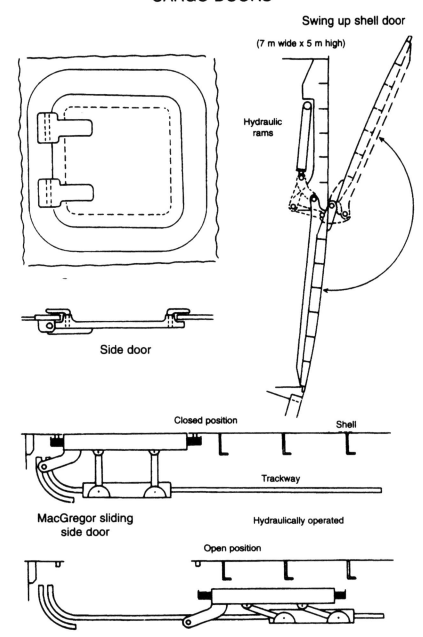

Swing up shell door

(7 m wide x 5 m high)

Hydraulic rams

Side door

Closed position

Shell

Trackway

MacGregor sliding side door

Hydraulically operated

Open position

Stern doors and ramps

The illustrations show the arrangements at the bow and stern of roll on/roll off ships.

The ship illustration on page 93 permits through movement of vehicles. Many roll on/roll off ships only have stern doors and ramps. The arrangements vary depending on the nature of the service being operated.

When using the bow doors it is necessary for the bow to run into a specially designed fender. Watertight closure at the bow is usually provided by the stern ramp in its raised position, or a separate door inside the bow door or visor. The bow door or visor has a spray type seal.

At the stern the ramp when in the raised position usually forms the watertight closure as shown.

The ramps and doors are normally hydraulically operated and cleated in position.

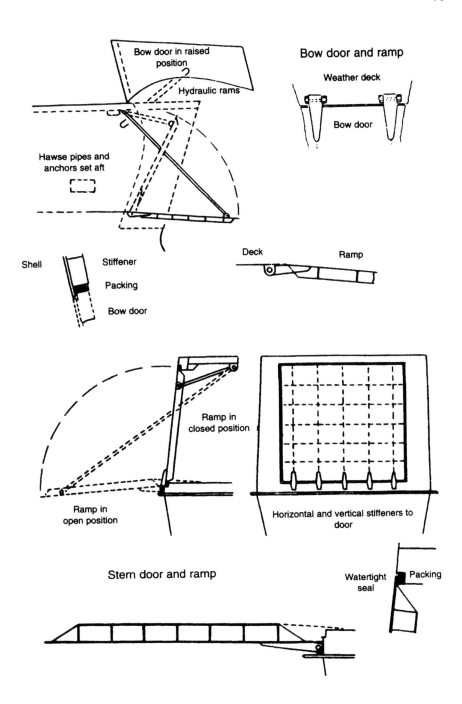

Bow door in raised position

Hydraulic rams

Bow door and ramp

Weather deck

Bow door

Hawse pipes and anchors set aft

Shell

Stiffener

Packing

Bow door

Deck

Ramp

Ramp in closed position

Ramp in open position

Horizontal and vertical stiffeners to door

Stern door and ramp

Watertight seal

Packing

Bulwarks and guard rails

Bulwarks and guard rails are fitted for the safety of the crew and play no part in the structural strength of the ship. Requirements for these are prescribed by the International Load Line Convention.

They are to be at least 1 m high and plated bulwarks are to be stiffened by a strong top rail section and supported by stays from the deck. Spacing of the stays is not to exceed 1.2 m on the forecastle decks of tankers and bulk carriers with reduced freeboards. On other decks and on other ships the spacing is not to exceed 1.83 m. The thickness of bulwarks is increased in way of mooring fittings, eye plates etc.

Where bulwarks form wells ample provision is to be made for freeing the decks of water as rapidly as possible. Bulwarks must therefore be provided with 'freeing ports', the minimum area of which is specified by the Convention and depends on the length of well. The lower edge of the freeing port is to be as near to the deck as possible and the openings are to be protected by rails spaced approximately 230 mm apart. If hinged doors or shutters are fitted to freeing ports they must have ample clearance to prevent jamming and the pins or bearings are to be of non-corrodible material.

If the construction of bulwarks was made integral with the sheerstrake then the light plating of the bulwarks (being further from the neutral axis) would be subjected to considerable stress with the subsequent fracture. This would create a notch at the sheerstrake and might give rise to a serious structural failure. A 'floating' bulwark, clear of the sheerstrake, is illustrated. Bulwarks are not to be welded to the top edge of the sheerstrake within 0.5L amidships.

Where guard rails are fitted they are to consist of courses of rails supported by stanchions efficiently secured to the deck. The opening between the lowest course of rails and the deck is not to exceed 230 mm in height and above that course openings are not to exceed 380 mm in height. Where the ship has a rounded gunwale the stanchions are to be secured at the perimeter of the flat of the deck.

Scuppers

Sufficient scuppers are fitted in all decks to give effective drainage. Those draining open decks above the freeboard deck are led directly overboard (see illustration). Scuppers from decks below the freeboard deck are led to the bilges or they may be led overboard if above the waterline and fitted with appropriate non-return valves.

BULWARKS

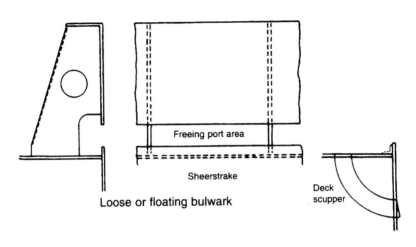

Freeing port area

Sheerstrake

Loose or floating bulwark

Deck scupper

Rails and freeing ports

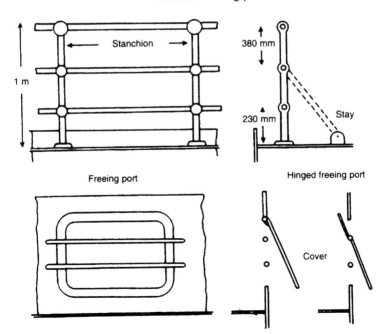

Stanchion

1 m

380 mm

230 mm

Stay

Freeing port

Hinged freeing port

Cover

Ventilators, air and sounding pipes

Ventilators are necessary to give adequate air circulation to under deck spaces, accommodation and tanks. The coamings of ventilators are to have a minimum height above the weather deck of 900 mm in Position 1 and 760 mm in Position 2 (see page 80). Where the coamings exceed 900 mm in height they are to be specially stayed.

All ventilators are to be provided with means of closing unless the height of the coaming exceeds 4.5 m in Position 1 and 2.3 m in Position 2.

Special care is to be taken when designing and positioning ventilator openings, particularly in regions of high stress concentration.

Mushroom, gooseneck and other minor ventilators are to be strongly constructed and efficiently secured to the deck.

Goose or swan neck type ventilators are mainly used for air pipes to tanks. The height (H) shown is not to be less than 760 mm on the freeboard deck and 450 mm on the superstructure deck. Air pipes are to be fitted at the opposite end of the tank to that at which the filling pipe is placed and/or at the highest point of the tank.

Sounding pipes are to be as straight as possible and to have a bore not less than 32 mm. Where a sounding pipe passes through a refrigerated compartment where temperatures may be 0°C or below the bore is not to be less than 65 mm. Striking plates of adequate thickness and size are to be fitted under open-ended sounding pipes.

VENTILATORS

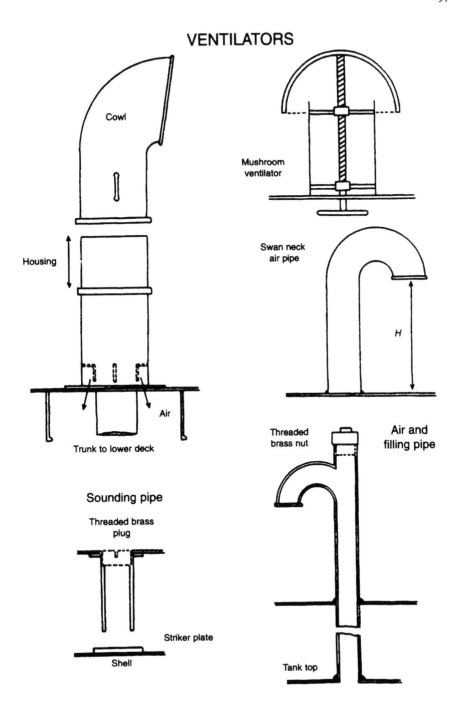

Cowl

Housing

Trunk to lower deck

Air

Mushroom ventilator

Swan neck air pipe

H

Sounding pipe

Threaded brass plug

Striker plate

Shell

Threaded brass nut

Air and filling pipe

Tank top

Mast and derrick posts

These are subjected to stresses when the derricks are lifting weights and their scantlings will depend on their height and the safe working load of the derricks

Stresses are high at the points of attachment so that additional support or stiffening is required at the heel, where the post passes through a deck and in way of derrick fittings. Heavier insert plates are welded into the deck at the heel and where the post passes through a deck. The heel requires support from below for vertical loads carried by the post or mast, which are ideally transferred to a bulkhead below the main deck. The mast or posts may be stiffened by welded pads or doublers in way of the attachment of derrick fittings.

The accompanying illustration shows a large bracket fitted directly below the heel which transfers the thrust of the mast to the adjacent bulkhead and adjoining structure.

The section through the shaft tunnel shown supports a mast or pillar. The scantlings in this area would be heavier than for the rest of the tunnel.

MAST STEPS

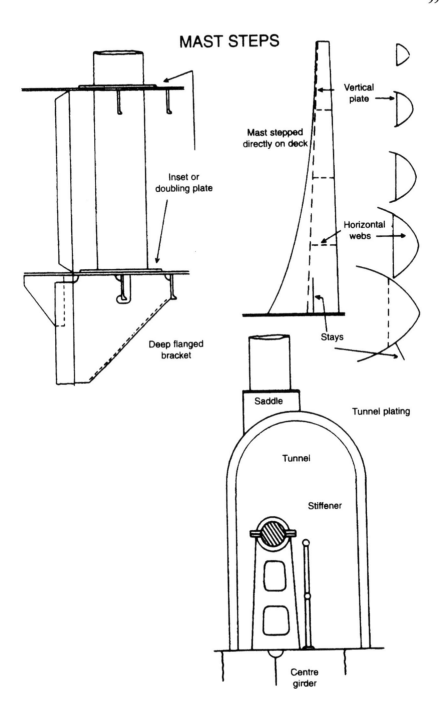

Inset or
doubling plate

Deep flanged
bracket

Mast stepped
directly on deck

Vertical
plate

Horizontal
webs

Stays

Saddle

Tunnel

Stiffener

Tunnel plating

Centre
girder

Patent heavy lift derricks

Many ships are now fitted with patent Hallen, Stulcken and similar heavy lift derricks and lifting gear. An outline arrangement of a Stulcken derrick is illustrated.

Derrick goosenecks, illustrated, allow a derrick to be swung in a horizontal, vertical or oblique plane. Means must be provided to prevent pins from lifting and, as shown, a check nut is fitted for this purpose.

Deck cranes

A large number of ships are fitted with cranes instead of conventional derricks for the handling of cargo.

They have a better overall performance, 360° of rotation, 100% flexibility and are faster in the handling of cargo.

The illustration shows a typical deck crane in use aboard a cargo ship.

OUTLINE STULCKEN DERRICK

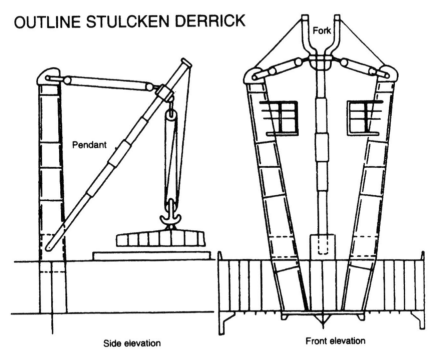

Side elevation

Front elevation

DECK CRANE

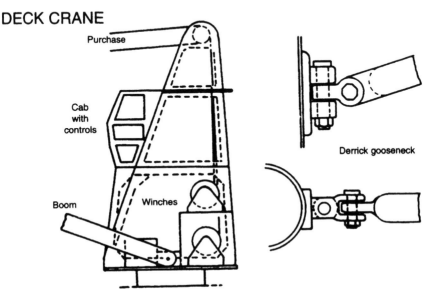

Rotating base

Deck fittings

Local stresses are frequently found in the way of deck fittings and steps are taken to strengthen the plating in their vicinity.

MOORING BITTS are attached by welding to the deck or bolting to welded stools on the deck.

PANAMA LEADS may be either fitted to the deck or in the bulwark as illustrated. The bulwark plating is increased in thickness in way of the lead.

FAIRLEADS are of many types, a roller type being illustrated. They may be attached directly to the deck or to the deck and bulwarks.

'OLD MAN' OR 'DEAD MAN' is a type of fairlead used to prevent chafing and to give a direct lead of a mooring line to the windlass, winch or capstan.

MULTI ANGLE FAIRLEADS reduce the number of fairleads required for mooring, warping, locking and docking operations. They are typical of the type required for operations in the St. Lawrence Seaway.

EYEBOLTS AND RINGBOLTS and other attachments for the use with derrick gear etc. are welded directly to welded pads on the deck, posts, bulwarks etc.

WINCHES AND WINDLASSES: a lot of stress is caused in the vicinity of deck machinery and they are usually securely bolted to a base which is welded on the deck. The deck will have a heavier insert plate in way of the item and some additional structure may be fitted below, including pillars, to distribute the load over a wider area of structure.

DECK FITTINGS

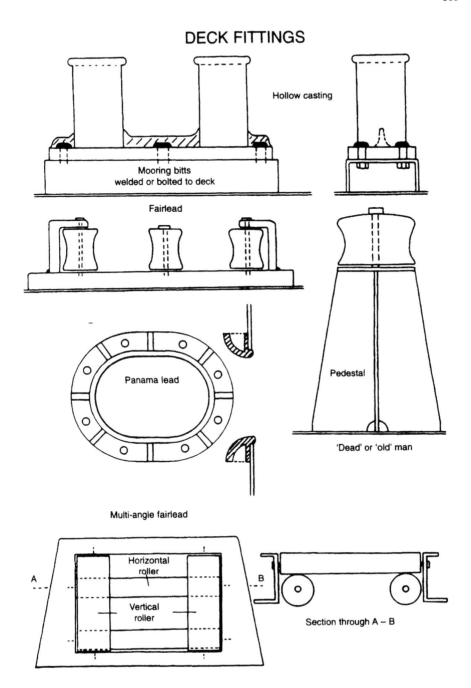

Hollow casting

Mooring bitts
welded or bolted to deck

Fairlead

Panama lead

Pedestal

'Dead' or 'old' man

Multi-angle fairlead

Horizontal
roller

Vertical
roller

A

B

Section through A – B

Engine rooms

The illustrations show the arrangement through an engine room that is situated amidships and one that is situated aft. A section through the aft engine room of a twin screw passenger ship is shown on page 105.

The location of the engine room is dependent on a number of factors such as the type of ship, number of screws, type of machinery etc. In cargo ships it ranges between aft and slightly aft of amidships, whilst in passenger ships there is also a trend for the machinery to be aft of midships. Tankers and bulk carriers always have the machinery aft.

The main engine seating is to be integral with the double bottom structure, the tank top plating in way of the seating being substantially increased in thickness.

Boiler bearers are to be of substantial construction, but, in order to allow for expansion the connection of the boilers to the bearers is not rigid.

Adequate transverse stiffening is required throughout the double bottom, with solid plate floors at every frame, and additional side girders to give the necessary support and strength.

Additional transverse strengthening is to be provided by means of web frames and strong beams with suitable pillaring or other arrangements. The webs are to be spaced not more than five frame spaces apart and are to have a depth of at least two and one-half times the depth of the normal frame.

Where the machinery is aft the double bottoms are to be transversely framed. Webs as above are to be fitted whether the side framing is transverse or longitudinal.

In the machinery spaces two means of escape, one of which may be a watertight door, are to be provided. There must be two means of communication between the bridge and a engine room or control room.

ENGINE ROOM

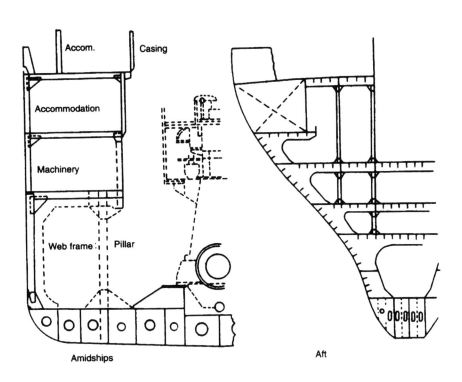

Accom.   Casing

Accommodation

Machinery

Web frame   Pillar

Amidships

Aft

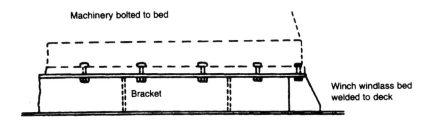

Machinery bolted to bed

Bracket

Winch windlass bed
welded to deck

Pumping and piping arrangements

Bilge pumping and piping arrangements in every cargo ship should, in general, be capable of discharging water from any compartment when the ship is on an even keel or listed not more than 5° either way. In the machinery spaces, additional arrangements are required so that any water may be discharged through at least two bilge suctions, one connected to the bilge main and one to an independent pump or ejector. An emergency suction must also be provided with a connection to the main circulating water pump in the case of a steam ship or to the main cooling water pump in the case of a motor ship.

Bilge and ballast lines may be constructed of cast iron, steel, copper or other approved material. Heat-sensitive materials such as lead must not be used. The size of bilge lines is determined by a formula depending on the main dimensions of the ship but is never to be less than 50 mm bore. Provision for expansion is made in the form of expansion bends or glands.

Screw down non-return valves must be provided on bilge lines, and mud boxes are provided on the suction bilge lines from machinery spaces to protect the pumps.

Strum boxes are fitted to the ends of suction bilge lines from spaces outside the machinery spaces. These strum boxes have perforations not more than 10 mm diameter with a total area of at least twice that required for the bore of the suction pipe. It should be possible to clear the strums without breaking any joint in a suction pipe.

Not less than two power-operated bilge pumps are to be provided for cargo ships, one of which may be operated by the main engines.

An arrangement of piping and pumps in an engine room is shown diagrammatically opposite. The bilge system is only designed to discharge overboard, but the ballast system is capable of discharging overboard, running up tanks by gravity and if necessary, pumping up tanks.

The forepeak tank has to have a screw down valve operated from above the bulkhead deck. This valve is to be inside the tank.

PUMPING AND PIPING

Bilge and ballast lines – diagram

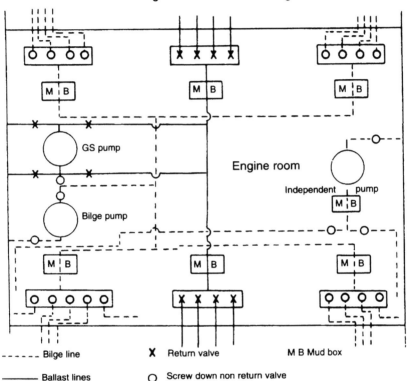

- - - - - Bilge line   **X** Return valve   M B Mud box

———— Ballast lines   ○ Screw down non return valve

Strum box

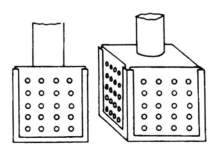

A general 'dry' cargo ship

The ship illustrated has longitudinal framing at the decks and in the double bottoms, transverse framing at the sides. This arrangement is that recommended in the Rules, see page 50.

To provide the necessary degree of transverse strength, transverses are fitted at the decks, see page 50, and plate floors are fitted in the double bottoms, see page 42.

Longitudinal framing is not usual at the sides of general cargo ships since this would necessitate the fitting of deep transverses 3.8 m apart (to provide transverse strength) and would give rise to a large amount of broken stowage.

In cargo ships a ceiling is to be laid over the bilges and under hatchways; the ceiling over the bilges being arranged with portable sections that are easily removable. Where no ceiling is fitted under hatchways the tank top plating thickness is increased by 2 mm. A wood ceiling laid in the square of a hatch is to be not less than 65 mm thick and should be laid directly on the tank top plating, being embedded in a suitable composition, or laid on battens providing a clear space for drainage.

- Where the tank top extends to the ship's side, as illustrated, bilge wells are fitted of not less than 0.17 cubic metres capacity.

Cargo battens are to be fitted in the holds from above the upper part of the bilge to the under side of the beam knees and in all cargo spaces in tween decks and superstructures. Wood cargo battens are to be 50 mm in thickness, the clear space between rows not exceeding 230 mm.

GENERAL CARGO SHIP

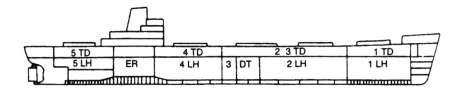

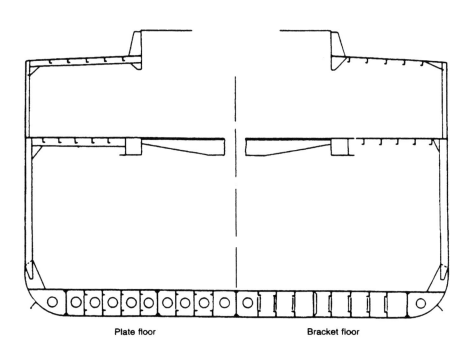

Plate floor   Bracket floor

Refrigerated ships

These ships are used for the carriage of cargoes which would deteriorate at ordinary hold temperatures.

To facilitate the fixing of insulation, transverse framing at the decks may be fitted in preference to longitudinal framing and transverses. Longitudinal framing is still fitted in the double bottom and may be preferred for the upper (strength) deck.

The holds and tween decks are insulated by packing an insulating material (fibreglass, silicate of cotton, slab or granulated cork) between the frames. This is held in place by wood sheathing or, galvanized steel or aluminium alloy sheeting. The latter is that most often used in modern refrigerated ships. The deckhead is insulated in a similar manner. Tank top insulation (illustration on page 45) is slightly different as the insulation has to be load bearing, also a 50 mm air gap or 5 mm of an oil-resisting material has to be provided in way of the crown of oil tanks. Tank top insulation in way of the square of the hatch requires sheathing.

Portable hatchways are insulated by insulated beams and plugs as illustrated. Note the shape of the plug hatches to ensure a tight fit. Patent steel covers may be filled with an insulating material. Other openings must be plugged and masts, pillars and other structure within the refrigerated space must be insulated.

There is to be provision for drainage from insulated spaces and this is effected by brine-sealed traps. The pipe from the lower hold must have a non-return trap which prevents odours from the bilges reaching the cargo spaces as well as preventing the bilges from freezing.

To cool the spaces, cold air is circulated through ducts in the cargo spaces. The air is drawn over a 'battery' of pipes through which cold brine (cooled by the expansion of a suitable refrigerant) circulates before being blown into the holds and chambers; fans etc. being fitted for this purpose.

Surveys are held before each cargo is loaded, when the cargo surveyor checks the cargo spaces for cleanliness and sound insulation and the temperatures are noted. The refrigeration machinery is examined under working conditions.

CROSS-SECTION THROUGH
REFRIGERATED VESSEL

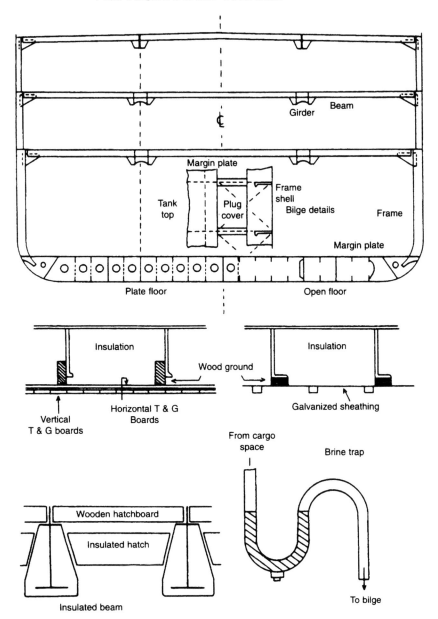

Dry cargo/container ship

The ship illustrated is capable of loading break bulk cargo, pallets, and containers. The upper deck is suitably stiffened, and hatches constructed accordingly, for the carriage of containers on deck.

Side doors are fitted at the upper tween deck level to facilitate loading direct from the quay to the ship.

Sliding bulkhead doors are fitted, connecting the holds in the tween decks and upper tween decks. They are pneumatically operated.

Flush fitting hatch covers are fitted in the lower decks to permit the use of fork lift trucks.

The height of the ship has been carefully designed so as to permit the stowage of five tiers of containers.

Large centreline pillars are fitted in conjunction with a centreline girder and cantilevers at the ship's sides. The spacing of pillars and cantilevers is as illustrated in the profile.

The holds are sealed by watertight electrically driven hatch covers on the weather deck, stowage for the covers being arranged at each end of the hatch.

Ships of this type are frequently fitted with electrically driven deck cranes to serve the various holds.

DRY CARGO – CONTAINER CARRIER

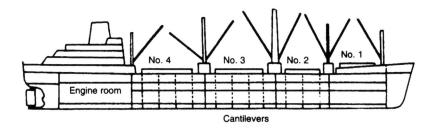

Cantilevers

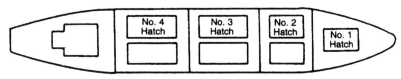

Main deck hatch arrangement

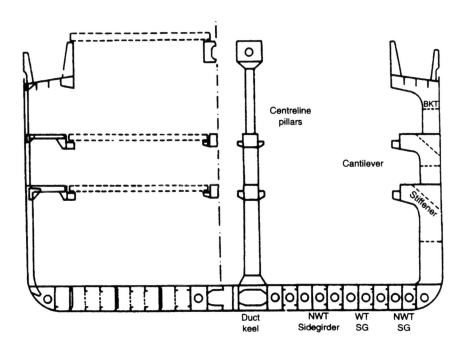

Oil tankers

Oil tankers are often divided into two categories, the smaller product carriers, carrying refined oil products, and the large crude oil carriers. The former have a greater number of tanks and more complicated pumping arrangements because of the variety of products carried.

Both types of ship are single flush deck ships with longitudinal and transverse bulkheads forming the tank space. The arrangement within the tank space is dictated by the requirements of the MARPOL Convention (see page 11). In particular the requirements for tank length, not to exceed $0.2L$ where two or more longitudinal bulkheads are fitted, and the provision of a protective location for segregated ballast tanks in this region. From 1994 new tankers of 5000 tonnes deadweight or more are required to have a double hull construction in way of the tank spaces or other equivalent means of protection against oil pollution.

Tankers have a longitudinally framed bottom shell and deck throughout the tank spaces. In tankers of not more than 150 m in length the side shell may be transversely framed and the longitudinal bulkheads may be vertically stiffened. Such framing has been referred to as 'combined' framing. In larger tankers all framing is longitudinal and is supported by large transverses 3 m to 5 m apart, depending on the ship's size.

At the ship's centreline deep girders are fitted at the deck and bottom (the bottom girder is often referred to as a 'docking girder') which align with a deep vertical centreline web on the transverse bulkhead.

In the transversely framed wing tank spaces horizontal stringers with cross ties are fitted to support the frames and vertical stiffeners. Similarly in longitudinally framed wing tanks two or three cross ties connect the deep webs supporting the longitudinals at the ship's side and longitudinal bulkhead.

Bulkheads may be corrugated rather than stiffened, with the corrugations arranged vertically on transverse bulkheads and horizontally on longitudinal bulkheads. Transverse bulkhead stiffeners are supported by horizontal girders which align with those on any transversely framed sides and longitudinal bulkheads.

In double-hulled tankers the double bottom has longitudinal framing with supporting plate and bracket transverse floors similar to cargo ships. The double hull side space and any hopper tanks are also longitudinally framed with support from transverses aligned with the bottom floors, bulkheads etc. (see page 117).

At the ends of the tank spaces cofferdams with two adjacent oiltight bulkheads are to be fitted. A pump room aft or a ballast tank forward may be accepted in lieu of a cofferdam.

The machinery space and aft peak has a transversely framed double bottom with transverse or longitudinally framed sides and deck, the construction being similar to that of other cargo ships.

The fore end may be transversely or longitudinally framed and is similar to that for other ships with panting arrangements etc.

The openings for oiltight hatchways are kept as small as possible and the corners are well rounded. Coamings should be at least 600 mm high and suitably fastened (gastight) with steel or other approved material covers fitted.

TANKER – COMBINED SYSTEM

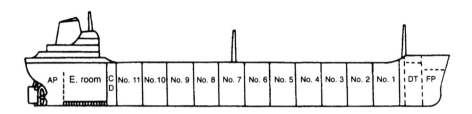

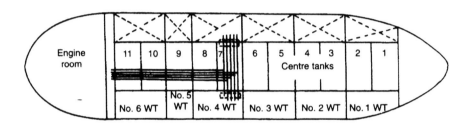

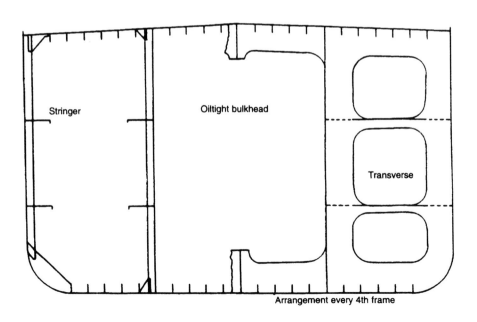

Arrangement every 4th frame

V.L.C.C. TANKER
Longitudinal system

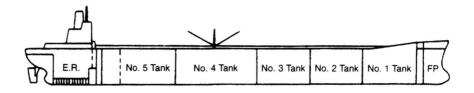

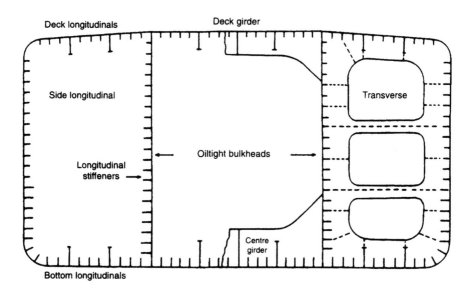

DOUBLE-HULLED
PRODUCT CARRIER
(13 000 TONNES DWT)

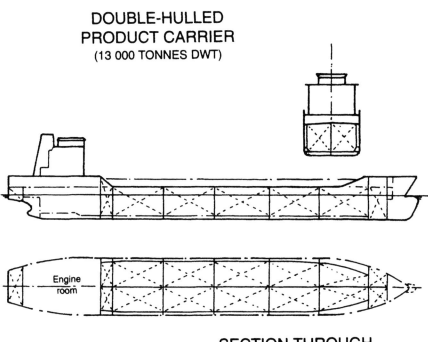

SECTION THROUGH
DOUBLE-HULLED VLCC
(290 000 TONNES DWT)

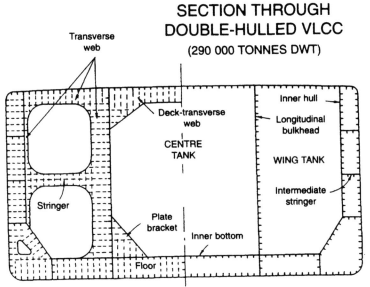

Bulk carrier

This type of ship is designed to load a maximum deadweight of any type of bulk cargo, from heavy ore to light grain. The ship illustrated has been designed to carry bulk sugar as the main commodity.

The ship is constructed on the combined framing system, having longitudinal framing in the double bottoms, bottom of wing tanks and at the deck, with transverse framing being fitted at the sides. Transverse webs are fitted in the wing tanks at intervals of 3.4 m, side stringers being fitted at approximately one-third and two-thirds the depth of the tanks.

This ship has two longitudinal watertight bulkheads which permits the ship to have a minimum freeboard under the International Load Line Convention, equivalent to that for a tanker.

The wing tanks may be used for the carriage of grain, bulk cargo or water ballast.

BULK CARRIER

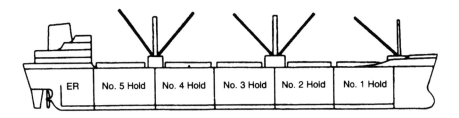

Grain, bulk cargo, water ballast carried in wing tanks

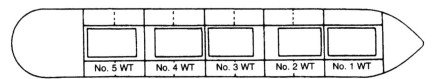

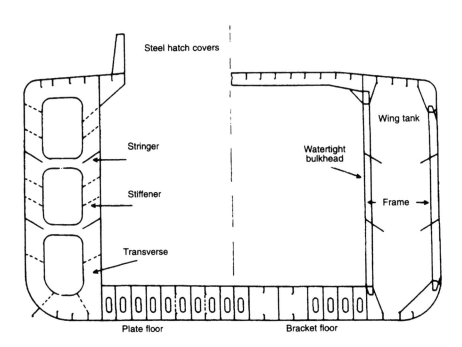

A modern collier or bulk carrier

This type of ship is designed for rapid loading and discharging of cargo. To achieve this, very wide, long and high hatchways are fitted. Hatch covers are made of steel with direct securing arrangements.

Topside tanks and hopper tanks (an extension of the double bottoms up the ships sides) are fitted to give adequate ballast capacity and thus adequate stability and draught when in ballast conditions.

The after deck is frequently raised to form what is known as a 'raised quarter deck' in order to give increased cargo capacity aft and thus prevent any tendency to trim by the head when fully loaded.

The mast may be telescopic in order to safely negotiate low bridges.

A profile and cross-section of a modern collier is illustrated opposite.

COLLIER

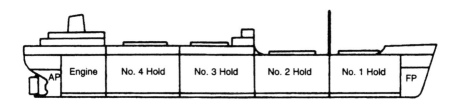

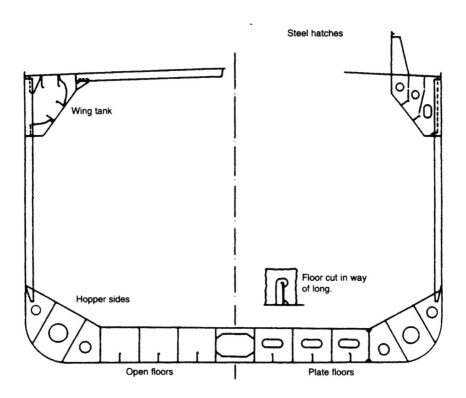

Bulk ore carriers

These are designed to carry a high density cargo, and the particular requirements of this trade are for the ship to have adequate strength for the heavy loading, a centre of gravity when fully loaded which is high enough to avoid undue roll 'stiffness', and arrangements to ensure maximum speed in loading and discharging.

The construction of such bulk carriers varies considerably, the type illustrated being the most common arrangement.

The ship illustrated has its cargo space divided into six self-trimming cargo holds and is so designed that full cargoes of grain may also be carried without the use of shifting boards. The wide hatches are fitted with hydraulically operated patent steel hatch covers.

Adequate ballast capacity is given by the tankage formed by the topside wing tanks and hopper tanks.

The ship shown has the following particulars, L = 160 m, B = 23 m, D = 13 m, draught = 9.4 m, Deadweight = 22 000 tonnes. About 10 000 tonnes of water ballast may be carried. No. 3 hold has been constructed as a deep tank.

The construction of this type of bulk carrier has been the subject of much debate and investigation following a high casualty rate during the 1980s. Whilst this debate still continues it is believed that local structural failures leading to the loss of watertight integrity of side shell plating, and subsequent progressive flooding following the collapse of watertight bulkheads, is the probable cause of many of these casualties. Changes to international requirements relating to the design and operation of these ships are still under consideration at IMO, but requirements have been put in place for enhanced surveys of bulk carriers, since maintenance of the structure was considered a significant factor. Also Rule requirements have been modified, in particular, that relating to side frame attachments.

ORE CARRIER

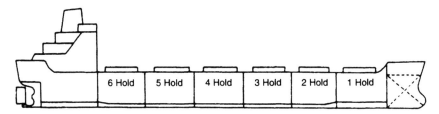

6 Hold | 5 Hold | 4 Hold | 3 Hold | 2 Hold | 1 Hold

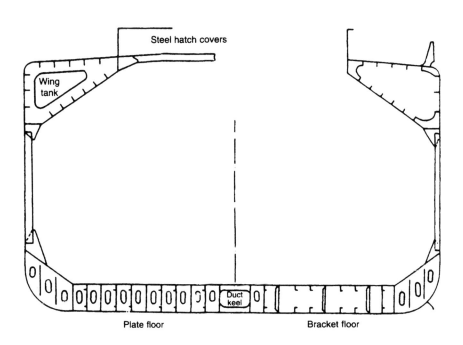

Steel hatch covers

Wing tank

Duct keel

Plate floor   Bracket floor

OBO (ore/bulk/oil) carrier

The major design characteristic peculiar to this type of ship as compared with an ore carrier is the double skin at the sides, having all the stiffening within the narrow wing tanks.

Advantages of the double skin are:

(1)   it makes for easier cleaning of the holds;
(2)   the inner skin reduces free surface in the large cargo holds; and
(3)   the clean ballast capacity of the ship is increased.

Transverse bulkheads are usually of the cofferdam type with all the stiffening in the cofferdam. Though easier to clean the holds there is a loss of cargo capacity.

There is usually a rise in the floor of the inner bottom which facilitates drainage to drain recesses or wells arranged on the centreline.

Hatch covers are of the side rolling type. The hatch breadth should be approximately 50% of the beam. The illustration on page 125 shows the type of hatch cover used, gas tightness being essential. Hydraulic operation with automatic battening down is a feature of these hatch covers.

The OBO carrier is technically different from the dedicated ore or oil carrier. The OBO carrier has a smaller deadweight compared with a bulk ore carrier or crude oil tanker of the same dimensions.

The flexibility of this type of carrier is of particular appeal to an owner who has a network of contracts covering the transportation of many commodities over a large area. An OBO carrier has the ability to switch readily from dry to liquid spaces.

OBO CARRIER

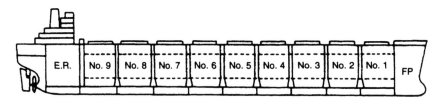

Double bulkhead between holds

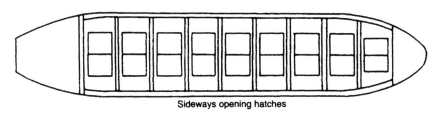

Sideways opening hatches

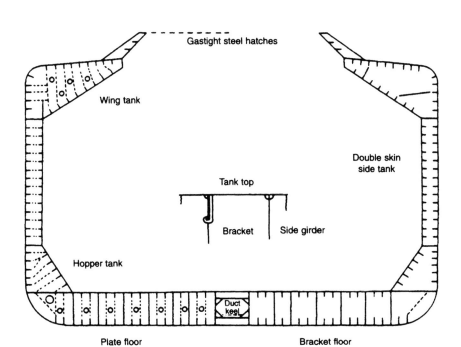

Liquid petroleum and natural gas carriers

The hull construction of this type of ship is often similar to that of an ore carrier, topside wing and hopper tanks being fitted for the carriage of water ballast.

Longitudinal or transverse framing may be used at the sides with an inner hull occasionally being fitted. Both arrangements are illustrated.

The cargo of liquefied gases is carried in independent tanks or membrane tanks. Membrane tanks are those which provide only containment for the liquid and rely on structural support from the adjacent ship structure. The non-pressurized tanks are rectangular or trapezoidal in section and may be independent or membrane tanks. Independent pressurized tanks are cylindrical. All such tanks are fitted with a dome extending through the deck giving access to the pipes, pumps and gauges.

LIQUID PETROLEUM GAS CARRIER

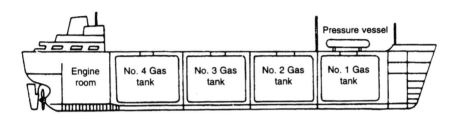

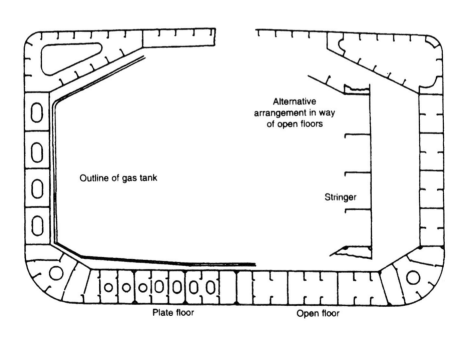

Independent tanks are located by supports on the double bottom with sufficient clearance for inspection of the tank. Provision must be made to ensure that the cargo tanks will not move when the ship is pitching and rolling in a seaway, also they must not be permitted to float should the hold become flooded when the tanks are empty. When the cargo is refrigerated provision is to be made for expansion and contraction of the tanks.

Liquefied gases are usually transported in one of three ways, namely:

(1)   under pressure at ambient temperature;
(2)   fully refrigerated at their boiling point; or
(3)   semi-refrigerated at reduced temperature and elevated pressure.

The way in which they are transported not only influences the type and construction of the tanks but also dictates the materials that can be used.

The illustration shows an independent tank installation for a refrigerated un-pressurized gas cargo. The tank shown consists of a structural inner skin or primary barrier supported by internal webs and stiffeners and a structural outer skin or secondary barrier connected to the inner skin by web plates. The primary and secondary barriers are often constructed of corrugated plating to obtain the necessary structural rigidity without stiffeners. The tank is insulated as illustrated.

All gas ships have void spaces in way of the tanks which are monitored for gas leaks and in many ships these are filled with an inert gas at a pressure slightly above atmospheric in order to keep out air and moisture.

LIQUID CARRIER

Shell Inner Primary
skin insulation

Inner membrane
of tank

Secondary   Outer
insulation   membrane

Shell Inner Outer
skin membrane

Inner
membrane

Stiffeners
to tank

Chock

Inert gas

Glassfibre

Insulation

Gas tank

Double skin   Longitudinal
stiffener

Web   Vertical
stiffener

Anti-roll chock

Tank

Supports

Anti-pitch chock   Deck

Tank

Collision chock

Supports

Container ships

The main object in the design of these ships is to carry the maximum number of containers within the designed length and breadth having regard to the form and structural arrangement.

The provision of adequate structural strength, given the large deck openings, is of prime importance. Longitudinal framing is used throughout the main body of the ship, transverse framing being used in the fore and after parts. These ships are built having a cellular construction at the sides. Strong longitudinal box girders are formed port and starboard by the upper deck, passageway flat, upper side shell and top of the inner hull shell. High tensile steels are frequently used for the upper deck and sheerstrake which are integral members of this structural box. In addition to providing longitudinal strength these box girders are designed to resist torsional stresses at the deck, given the lack of structural material resulting from the large deck openings. They are referred as a 'torsion box'.

The hatchway illustrated is divided into three sections, two long hatch girders being fitted. The girders are made continuous thus sharing the longitudinal bending strength and adding to the sectional modulus.

The carriage of containers above the main deck results in a high loading on the deck and the hatch covers which are strengthened to withstand this extra loading.

The container spaces are suited either for 12.20 m or 6.10 m units. A form of bulkhead is fitted at intervals of 14.70 m, centre to centre with watertight bulkheads being fitted as required by the Rules. The bulkheads provide resistance to racking.

The container guides and associated structures are designed to withstand dynamic (accelerating) forces due to rolling, pitching and heaving. The guide consists of angle bars typically 150 mm by 150 mm by 14 mm connected to vertical webs and adjoining structure, spaced 2.60 m apart. The bottom of the guides are bolted to brackets welded to the tank top and beams. The brackets are welded to doubling plates, 15mm thick, which are welded to the tank top.

CONTAINER SHIP

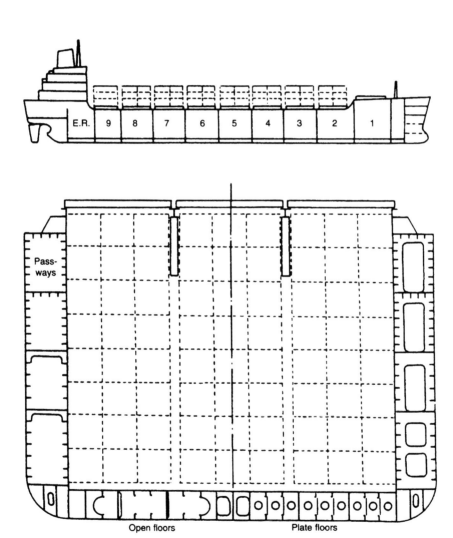

Passenger ships

The illustration opposite shows the profile and cross-sections of a twin screw passenger. ship. The cross-sections shown are the midship section, and in way of the engine room and an after compartment. The engine room is situated abaft amidships.

Any ship which carries more than 12 passengers is considered as a passenger ship under the provisions of the SOLAS convention (see page 11) and this convention has a major influence on the design and construction of such ships.

Passenger ships range from the large ocean liners and cruise ships with space for little or no cargo to shorter voyage passenger ships, many of which carry Roll on/Roll off cargoes and passenger vehicles. Most ocean going passenger ships are now cruise ships and the passenger ferry type dominates with a wide variety of ships including the increasing number of high speed craft.

In the construction of these ships the SOLAS requirements predominate in relation to subdivision, intact and damage stability and structural fire protection. Rule requirements take into account the global and local stress considerations of these ships, particularly in respect to the large superstructures fitted.

The superstructures are frequently constructed of aluminium alloys which, in addition to the reduction in weight, improves the stability.

Since the comfort of passengers is very important many of these ships are fitted with stabilizers, and bow thrusters are provided to assist manoeuvrability at low speeds.

The design of these ships is highly specialized, particularly the accommodation areas.

PASSENGER VESSEL

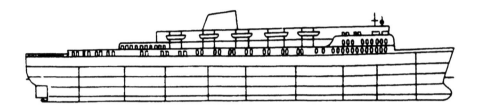

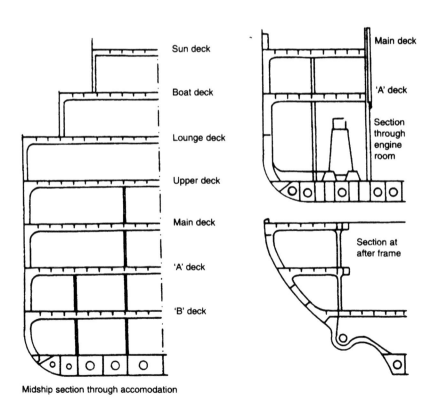

Sun deck

Boat deck

Lounge deck

Upper deck

Main deck

'A' deck

'B' deck

Midship section through accomodation

Main deck

'A' deck

Section
through
engine
room

Section at
after frame

Roll on/Roll off passenger ship

The illustration shows the profile and midship cross-section of a Roll on/Roll off passenger ship. Noticeable are the ramps and doors at the bow and stern to facilitate the loading and discharge of vehicles, these are detailed on page 93.

A feature of the vehicle deck is a clear deck uninterrupted by transverse bulkheads. Deck heights are to be sufficient to accommodate the various types of vehicles that are to be carried. In the ship illustrated the lower decks are used for cars and the upper decks for larger vehicles and trailers.

Transverse strength is maintained by fitting deep, closely spaced web frames in conjunction with deep beams. These may be fitted at every 4th frame, about 3 m apart.

The lower decks divided by watertight transverse bulkheads, have hydraulically operated and cleated sliding watertight doors to facilitate the movement of vehicles. These decks are reached by fixed or hydraulically operated ramps and lifts. Movable ramps have the advantage of permitting additional vehicle stowage space.

Some Roll on/Roll off ships have the stern door set at an angle to the ship's centre-line so that full Roll on/Roll off operations can be maintained alongside a normal quay.

RO-RO FERRY

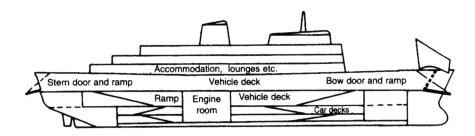

Accommodation, lounges etc.
Stern door and ramp — Vehicle deck — Bow door and ramp
Ramp | Engine room | Vehicle deck
Car decks

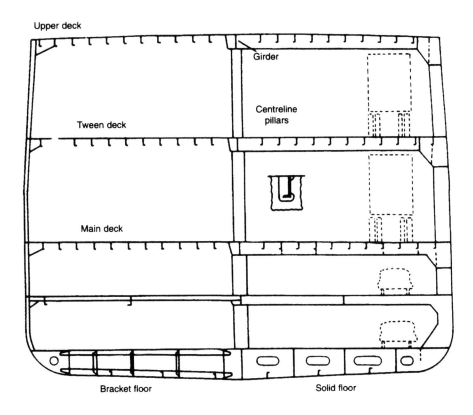

Upper deck

Girder

Centreline pillars

Tween deck

Main deck

Bracket floor — Solid floor

Index